Visions of Mars

Visions of Mars

*Essays on the Red Planet
in Fiction and Science*

Edited by
HOWARD V. HENDRIX,
GEORGE SLUSSER
and ERIC S. RABKIN

McFarland & Company, Inc., Publishers
Jefferson, North Carolina, and London

LIBRARY OF CONGRESS CATALOGUING-IN-PUBLICATION DATA

Visions of Mars : essays on the red planet in fiction and science /
edited by Howard V. Hendrix, George Slusser, and Eric S. Rabkin.
p.   cm.
Includes bibliographical references and index.

ISBN 978-0-7864-5914-8
softcover : 50# alkaline paper ∞

1. Mars (Planet)—In literature.   2. Science fiction—History and criticism.
3. Mars (Planet)—Exploration.   I. Hendrix, Howard V., 1959–
II. Slusser, George Edgar.   III. Rabkin, Eric S.
PN3433.6.V57   2011   809'.93329923—dc22   2010053294

British Library cataloguing data are available

© 2011 Howard V. Hendrix, George Slusser, and Eric Rabkin. All rights reserved

*No part of this book may be reproduced or transmitted in any form
or by any means, electronic or mechanical, including photocopying
or recording, or by any information storage and retrieval system,
without permission in writing from the publisher.*

On the cover: Mars as seen from the Viking 1 Orbiter on July 11, 1976 (NASA);
(inset) cover art by Frank Schoonover from the 1917 edition of *A Princess of Mars*

Manufactured in the United States of America

*McFarland & Company, Inc., Publishers
Box 611, Jefferson, North Carolina 28640
www.mcfarlandpub.com*

Table of Contents

Preface: Science, Fiction, and the Red Planet
    George Slusser, Howard V. Hendrix, Eric S. Rabkin 1

Introduction: The Martian in the Mirror
    Howard V. Hendrix 9

One: Approaching Mars

Mars of Science, Mars of Dreams
    Joseph D. Miller 17

Where Is Verne's Mars?
    Terry Harpold 29

Rosny's Mars
    George Slusser 36

Two: The Uses of Mars

Dibs on the Red Star: The Bolsheviks and Mars in the Russian Literature of the Early Twentieth Century
    Ekaterina Yudina 51

The Martians Among Us: Wells and the Strugatskys
    George Slusser 56

Savagery on Mars: Representations of the Primitive in Brackett and Burroughs
    Dianne Newell and Victoria Lamont 73

The (In)Significance of Mars in the 1930s
    John W. Huntington 80

Spawn of "Micromégas": Views of Mars in 1950s France
    BRADFORD LYAU 86

Is Mars Heaven? *The Martian Chronicles*, *Fahrenheit 451* and
Ray Bradbury's Landscape of Longing
    ERIC S. RABKIN 95

Re-Presenting Mars: Bradbury's Martian Stories in Media Adaptation
    PHIL NICHOLS 105

Robert A. Heinlein and the Red Planet
    DAVID CLAYTON 118

Business as Usual: Philip K. Dick's Mars
    JORGE MARTINS ROSA 130

Kim Stanley Robinson: From *Icehenge* to *Blue Mars*
    CHRISTOPHER PALMER 139

Martian Musings and the Miraculous Conjunction
    KIM STANLEY ROBINSON 146

Chronicling Martians
    SHA LABARE 152

Three: Science and Fictional Mars

Mars as Cultural Mirror: Martian Fictions in the Early Space Age
    ROBERT CROSSLEY 165

Beyond Goldilocks and Matthew Arnold: Interplanetary Triage,
Extremophilia, and the Outer Limits of Life in the Inner Solar System
    HOWARD V. HENDRIX 175

*Appendix 1—To Write the Dream in the Center of Science: Mars and
the Science Fiction Heritage: A Dialogue Between Ray Bradbury
and Frederik Pohl (George Slusser, Moderator) (May 2008)* 185

*Appendix 2—The Extreme Edge of Mars Today: A Panel
Discussion with David Hartwell, Geoffrey Landis, Larry Niven, and
Mary Turzillo, Moderator (May 2008)* 190

About the Contributors 207

Index 211

PREFACE
Science, Fiction, and the Red Planet
George Slusser, Howard V. Hendrix, Eric S. Rabkin

One might imagine that, in the long road from princesses on Mars to the bare hope we have today for possible microbial life on the Red Planet, little interest would remain in placing stories there. A perusal of the web, however, proves this is dead wrong. There is more life on Mars than ever. One website mentions, for example, *The Mammoth e-book of Mindblowing Mars SF* (2009) which features 20 new short stories, flash fictions and poems, all set on Mars, by such SF writers as Kage Baker, Mary Turzillo, Patricia Stewart, Liz Williams, and others. This same website announces the following Mars projects: a graphic novel adaptation of *The Martian Chronicles* (2010); "The Waters of Mars," a *Dr. Who* TV special on the BBC (2010); *War of the Worlds*, a Goliath animated film (2010); *Life on Mars: Tales from the New Frontier*, a story anthology by Jonathan Strahan (2010); *The Lost Hieroglyph: A Brackett and Burroughs Adventure* (2010), and finally a Disney/Pixar film version of *John Carter of Mars* (2011).

These new Mars adventures are merely the latest visions in the long dream we trace in this book. The continuing presence of such fictions points to our need to understand where the fascination with Mars came from, and the nature and importance of this Martian odyssey in the Western world. The story of that fascination is closely related to science and its successive visions of Mars—from Huygens and Percival Lowell to the Mars Rovers. Mars has fascinated writers from many places and times, and this book deals with early writers, including Voltaire and Verne and their French followers; Wells's classic *War of the Worlds*; Russia's Mars in Bogdanov, the Mars of the Strugatskys; Burroughs, Rosny and Weinbaum; and into and through the early Golden Age of science fiction. *Visions of Mars* also focuses on the High Age of Mars—the Mars of Bradbury, Heinlein, Asimov, Dick—then moves to examine the post–*Mariner* Mars: the Dead

Planet, briefly brought to life in the 1990s by Kim Stanley Robinson, only to return to the hostile landscapes of Geoffrey Landis. Although no book short of a grand Martian encyclopedia can hope to even touch on all that has been said or written about Mars, here you will nonetheless find the vision of science fiction's Mars detailed in much of its richness, from Mars as object of scientific inquiry in "hard" science fiction, to the Mars-as-rich-metaphor in works ranging from Bradbury's *Martian Chronicles* to Gerrold's *The Martian Child*.

Howard V. Hendrix's introduction, "The Martian in the Mirror," offers a personal meditation on the Martian odyssey of science fiction. He sees the vision of Mars this book proposes as a vast compendium of theories, ideas, and fictions about another world. Whenever we look at Mars, we are looking at the mirror of our own cultural dreams and concerns.

The essays that follow are divided in three sections: (1) "Approaching Mars," the short opening section, offers a brief history of the scientific discoveries that have periodically brought Mars to our attention, and an appreciation of how Jules Verne, arguably the first science fiction writer, viewed Mars only as an occasional venue for adventure and discovery; (2) "The Uses of Mars," the main body of the text, offers individual examples of how writers beginning with H.G. Wells have used Mars to project their unique visions of humanity's future; (3) "Science and Fictional Mars," the short concluding section, presents two essays that broadly analyze the Martian phenomenon in science fiction: Robert Crossley's discussion of the radical change wrought by the *Mariner 9* photos, and Howard V. Hendrix's sweeping analysis of the nexus of planets, persons, and pathogens, and the triage relation across the history of the science fiction genre between and among Venus (too hostile to life to be worth "treating"), Earth (abundantly alive and therefore in no need of "treatment"), and Mars (worth "treating" because of its potential for supporting life).

Approaching Mars

In "Mars of Science, Mars of Dreams," Joseph D. Miller, a professional life scientist and a science fiction critic, explores Mars as "a Rorschach test for the imaginative since at least the time of the Italian astronomer Giovanni Schiaparelli. Schiaparelli, in 1877, observed what he believed to be straight lines (*canali*, or channels) in equatorial regions of Mars." Miller charts the development of both our knowledge of Mars and our certainty about that knowledge, indicating how exemplary fiction writers from Wells to the present exploit and sometimes distort the Mars known at any given moment to science. Once he brings his reader to an understanding of our current knowledge of Mars, Miller concludes with suggestions about the sorts of imaginative projections today's science makes possible.

In "Where Is Verne's Mars?" Terry Harpold begins with an examination of the relative paucity of the occurrence of Mars in Verne's fiction, linking it to Verne's almost obsessive emphasis on circuits of movement and the possibility that Verne had little need of Mars for demonstrating this favored literary mechanism, since the poles, tropics, and cities of Earth already provided him with space enough for such demonstration. Harpold then takes us beyond those bounds, however, by suggesting the ways Verne's work can serve as a model for reading modern fictions of Mars.

George Slusser's "Rosny's Mars" discusses J.H. Rosny's *Les Navigateurs de l'Infini (Navigators of the Infinite)* (1927). Rosny, a Belgian writer, penned evolutionary tales throughout his career, all set on Earth. His final SF novel, even though it projects an Earth-analog ecosystem on Mars, marks a significant departure from his earlier work. Not only do his astronauts, in their relations with the Martian humanoid species, affirm the anthropomorphism that Rosny rejected in his evolutionary novel *La Mort de la Terre (The Death of the Earth)* but they encounter, during the trip to Mars, the lure and terror of Pascal's infinite, toward which Mars is the first timid step. The novel takes French Mars stories in a wholly new direction, away from juvenile space opera and toward the space adventures of the American Golden Age, and the Mars of Anglophone figures like Clarke and Heinlein.

The Uses of Mars

Ekaterina Yudina's "Dibs on the Red Star: The Bolsheviks and Mars in the Russian Literature of the Early Twentieth Century" examines Russian and Soviet fascination with the Red Planet. Her touchstone text is Alexandr Bogdanov's prerevolutionary novel *Red Star* (1908). She sees most Russian Martian texts, even today, in dialogical relation to Bogdanov, and examines one prominent Soviet example, Alexei Tolstoy's *Aelita*. In Bogdanov's novel, Mars is a place of successful revolution. Tolstoy, however, significantly reverses the relationship: Now it is Earth that has achieved a successful socialist revolution while Mars is a dying planet with an oppressive society. The novel's answer to the question of whether or not Mars can be redeemed is ambiguous. Yudina contrasts Bogdanov's fascination with the color red (including the red blood of his later Institute for Blood Transfusions, an institution for aging Soviet leaders) with the essentially colorless nature of Tolstoy's Mars.

George Slusser's "The Martians Among Us: Wells and the Strugatskys" discusses the parallel roles of Mars and Wells in Russian and Soviet literature, leading to the confluence of British writer and Red Planet in the post–Stalin novels of the Strugatsky brothers. Not only is Mars a common theme of the Strugatskys, but Wells's *War of the Worlds* becomes a persistent subtext in their fiction, offering a vision of unregenerate human nature that undermines the official Marxist view of history as essentially progressive. Wells's Martians are already at work in the Strugatskys' "juvenile" *Space Apprentice* and haunt the pages of the Bradbury-like story chronicle *Noon: 22d Century*. The theme of invasion and reaction deepens in works from *Hard to Be a God* and *The Second Invasion from Mars* to *Ugly Swans* and *Roadside Picnic*. The Wells subtext becomes pervasive in *A Billion Years to the End of the World*, when Martians confront the Invisible Man.

Diane Newell and Victoria Lamont's "Savagery on Mars: Representations of the Primitive in Brackett and Burroughs" examines how Brackett's and Burroughs's different eras shaped their approaches to their heroes and their Martian worlds. The authors emphasize Mars as liminal space between the real and the fantastic, a place both imaginations and empires might colonize. At the very time the American frontier on Earth was closing, Burroughs and his immediate predecessors saw in Mars a world of adventure and romance where the colonial narrative was still very much viable — but, Brackett, in her Mars fictions of the 1940s and 1950s, saw the imaginary colonial world in ways complicated not only by ideology but also by ongoing scientific discovery (so our authors suggest).

John W. Huntington's "The (In)Significance of Mars in the 1930s" discusses the dialogue between the Mars fictions of that period, primarily those of C.S. Lewis and especially H.G. Wells's *Star Begotten* (1937), as a conversation not about politics but about the possibilities and limits of SF in a time of crisis. For Lewis, Mars is a utopia and its story a scientific romance. Wells's story breaks with his earlier *War of the Worlds*, by focusing, not on invasion from Mars, but on Mars as possible source of mutation-causing rays that turn humans into "Martians." The book's protagonist comes to believe he is a Martian, and is proud of the fact, reversing the surface Anglophilia of *War of the Worlds*. On the eve of World War II, Lewis redeems his faith in the humanist tradition. Wells's novel, on the contrary, is all about the effects of propaganda and the power of rumor and panic on the eve of invasion.

Bradford Lyau's "Spawn of 'Micromégas': Views of Mars in 1950s France" takes us from Voltaire's eighteenth-century short tale "Micromégas" to publisher Fleuve Noir's Anticipation line of science fiction novels in the 1950s, outlining for us in that journey the influence of the *conte philosophique* upon the development of French science fiction. The influence of the "philosophical tale," Lyau suggests, helps to explain French SF's continuing interest in the exploration of scales, limits, perspective, and a distinctly French tradition of examining ideas.

In "Is Mars Heaven? *The Martian Chronicles*, *Fahrenheit 451* and Ray Bradbury's Landscape of Longing," Eric S. Rabkin analyzes the materials — personal, cultural, and creative — that Ray Bradbury used and reused in establishing a compelling fairy-tale Mars in the world's imagination. Before Bradbury, most literate readers deemed "science fiction" a disreputable popular genre, at best a guilty pleasure beneath serious consideration. With Bradbury, the genre suddenly gained enormous respect. Yet established science fiction writers rightly saw Bradbury as anti-scientific, a charge he readily conceded. In two classic novels and a series of stories, this self-acknowledged writer not of science fiction but of fantasy transformed public opinion of the most crucial genre of our technological era in part by his gorgeous lyricism and in part by supporting a particularly American variety of Western culture's oldest utopian myth, a return to Eden not by intellect but by musical echoes of Poe, Twain, Melville, and Sherwood Anderson, all of whom inform Bradbury's small town ideal.

Phil Nichols reports, in "Re-Presenting Mars: Bradbury's Martian Stories in Media Adaptation," that while Ray Bradbury's prominent lyricism might make his work difficult to adapt to other media, in fact Bradbury is one of the most adapted authors of all time. Nichols focuses on radio, film, and television adaptations of three chapters of *The Martian Chronicles*, two — "The Third Expedition" (originally "Mars Is Heaven!") and "And the Moon Be Still as Bright") — set on Mars and one, "There Will Come Soft Rains," set on Earth. The analyses demonstrate that the more successful adaptations tend to ignore the Martian connection entirely, treating the first, for example, as a horror story. Bradbury himself is "forever playful" in his self-adaptations, never seeming to feel a need for literal translation. However, all the adaptations together confirm the source of Bradbury's power as a lyrical writer: in general, the radio adaptations, limited to the aural, succeed best.

In "Robert A. Heinlein and the Red Planet," David Clayton discusses Heinlein's third and most important of four Mars novels, *Stranger in a Strange Land*, as failing to effect the "end of human innocence," the new coming of the eternal Adam that Heinlein promised in his "future history." Clayton traces the "matter of Mars" back to Greene's *Planetomachia*, through later tragedies of wars between worlds, to conclude that Heinlein has turned this tradition into a "satyr play." In *Stranger*, technology reverts to astrology, the Earthman returns as a Martian spy, and a con man exploits "grokking" to found a sect based on sex, drugs, and power. Valentine Michael Smith, the Martian hope for new beginnings, is a false Christ martyred on the very mass media he uses to dupe a humanity that remains fallen.

Jorge Martins Rosa's "Business as Usual: Philip K. Dick's Mars" argues that Philip K. Dick is different from other Golden Age writers in his skepticism about planetary colonization. Mars may be a place of election for Dick's colonists, but it remains a place where humanity only replays the laws of profit and speculation that are destroying (or have already destroyed) Earth. Mars is clearly a place of false hope in novels like *The Simulacra* and *Do Androids Dream of Electric Sheep?* But in these novels, the Mars experience marks a desire for a homecoming, as the androids are very importantly "made on Earth." If there is to be any solution for problems on Earth, we need a temporary retreat to the forest. Mars, Rosa claims, is Dick's forest.

Christopher Palmer's "Kim Stanley Robinson: From *Icehenge* to *Blue Mars*," compares Robinson's first Mars-linked novel, *Icehenge*, with the 1990s trilogy of *Red Mars*, *Green Mars*, *Blue Mars*. Where Palmer sees the trilogy as energetic and problem-solving, he finds *Icehenge* to be hesitant, uncertain about both the "Mars question" generally and how humans should approach the question of terraforming specifically. Palmer's analysis demonstrates that these same uncertainties do persist in the Mars trilogy, but that they now serve the purpose of what he calls a "dialectic of transparency and non-transparency." Palmer reminds us that, within the body of Robinson's work — generally characterized by positive, problem-solving, energetic, extroverted and epiphanic narratives — there is also a narrative stream of matters unstable, unsolved, withheld, reflective, and amnesic. Palmer argues that the latter, darker stream does not detract from the former, but rather deepens and complicates Robinson's narrative of exploration and accomplishment.

Kim Stanley Robinson, in "Martian Musings and the Miraculous Conjunction," discusses two histories for the connections between science about Mars and science fiction about Mars: the larger twentieth-century popular and scientific context, and how that popular and scientific context influenced Robinson's own personal history as a science fiction writer writing about Mars. These macrocosmic and microcosmic histories dovetail in what Robinson (borrowing a term from Galileo) refers to as a "mirabili coniunctio," a "miraculous conjunction," in which his personal historical moment allowed him to interact with the evolving context of scientific understanding such that Robinson was "lucky" enough to write the first major trilogy of novels involving the terraforming of Mars.

In "Chronicling Martians," Sha LaBare gives us an ecocritical reading of the meaning of Mars, through which he attempts to dislodge us from our "comfort zone" of anthropomorphic and anthropocentric visions — not only of Mars but also of Earth. LaBare's essay argues that ecological thinking, which numbers human beings among — rather than apart from — all the world's (and potentially all the universe's) "critters," is vital to speculation in the science fiction mode.

Science and Fictional Mars

Robert Crossley's "Mars as Cultural Mirror: Martian Fictions in the Early Space Age" explores Mars fact and fiction from the early 1950s to the early 1970s. To Crossley, Early Space Age Martian fictions may be said to begin after Ray Bradbury's *The Martian Chronicles* (1950, the defining masterpiece of the "Romantic" period of Mars fiction), and end not long before Robinson's Mars trilogy of the 1990s made stunning use of the Mars that science now knows. During the Early Space Age particularly, fiction writers struggled against the constraints of scientists' evolving knowledge of Mars itself. Crossley's survey of the science and the fictions exposes the tensions between the earlier and emerging views, including along the way a powerful argument for the sober excellence of Ludek Pesek's *The Earth Is Near* (a novel little known in the English-speaking world), and the watershed significance of *Mars and the Mind of Man*, a select 1972 conference at Caltech in which scientists and writers awaited and then assessed the stunning, irrefutable photos *Mariner 9* returned from Mars.

Howard V. Hendrix, in the concluding essay, "Beyond Goldilocks and Matthew Arnold: Interplanetary Triage, Extremophilia, and the Outer Limits of Life in the Inner Solar System," examines *War of the Worlds* by H.G. Wells, *Last and First Men* by Olaf Stapledon, and Ray Bradbury's *Martian Chronicles* in the context of the evolving understanding of Mars' suitability (or lack thereof) for life generally, and for microbes specifically. That evolving context and its relationship to the evolving texts of the Mars literary tradition in science fiction mirror our current situation, in which the extreme and exotic environments we continue to search out on Earth — as stand-ins for another world — prepare us not only for a fuller, on-site understanding of that other world, but also work toward the preservation of the world in which we already dwell.

Appendices

Beyond the concluding essay, we present in the appendices the transcripts of two important presentations on Mars that occurred in May 2008. They are original to the present volume (and have been slightly revised to fit the format of this book).

The first is "'To Write a Dream in the Center of Science': Mars and the Science Fiction Heritage," a wide-ranging conversation between Frederik Pohl and Ray Bradbury examining the relationship between dreams and science in the histories of both science fiction (about Mars, and more generally) and of the space program and the human future in space. Bradbury and Pohl also discuss the roles of film and print media in our popular imaginings of both Mars and the future, and consider how Mars, as a scientific and cultural object of inquiry, has and might hereafter affect these authors' worlds and the writing we may all read.

In the second of the appendices, we offer "The Extreme Edge of Mars Today," a panel discussion featuring Nebula Award–winning science fiction writer and critic Mary Turzillo as moderator, and a panel of three Hugo winners with definite Martian "credentials": her husband, space scientist and science fiction writer Geoffrey Landis, legendary hard–SF writer Larry Niven, and genre-shaping critic and editor David Hartwell, also in May 2008.

Although the response to the reality of Mars in science and fiction is ever changing, we hope this volume provides not only thoroughgoing groundwork on a century and a half of the synergy between Mars research and Mars speculation but also suggests where that synergy might lead in the future.

INTRODUCTION
The Martian in the Mirror
Howard V. Hendrix

This book's title, *Visions of Mars*, plays upon two senses of the word "vision": something seen, and something dreamed. You may, if you like, think of the fictional treatments of Mars as "something dreamed," and the scientific understandings of Mars as "something seen," but it's not quite that simple. Just as dreams are in many ways dependent on daylight experiences for their source material, so too is the "something dreamed" in the fictional treatments of Mars often dependent on the "something seen" of science. Conversely, just as dreams also shape our daylight lives, so too have the "dreams" of the fictional treatments of Mars often shaped what scientists "see" on the Red Planet.

Like *The Martian Chronicles* by that soaring poet of the American dreamtime, Ray Bradbury, this volume is also to some extent a chronicle, a chronological record of historical events. Bradbury's book, its chapters headed with the month and year of each event the text is about to relate, at least fulfills the standard definition's first half—"a chronological record." The poetic conceit hidden in Bradbury's title, of course, is that the events it chronicles are fictional, not historical — at least not in any sense other than that the events are said to take place in an imagined "future history" of another world. Yet, curiously, time has shown that Bradbury's histories of times to come inform the very real worlds in which we live, work, and imagine. The times not only shape the dreams — the dreams also shape the times, especially if a dream-vision is, in some sense, already the chronicle of a fiction.

This volume of essays, panels, and addresses manipulates the concept of "dream vision" and "chronicle," but in a manner different from Bradbury's. The dates and times of the visions mentioned here do, in fact, chronicle historical matters. One can learn from *Visions of Mars* the answers to questions such as "What was the first published appearance of the theory of the 'dying Mars' in the scientific literature?" or "Into what literary-historical phases may the twentieth-century depictions of Mars in science fiction be categorized — and why?" In this volume, however, what are primarily "chronicled" are not dates and events so much as theories, ideas, and fictions —*visions*— of another world. A book which presents a scholarly overview of the ways in which Mars — its past, present, and future — has been envisioned in theory and fiction produced on Earth is probably, in its own way, almost as entangled with the imaginary as a book by a terrestrial author which purports to be a history of envisioned future events on Mars. Yet the question underlying both visions and chronicles of the Red Planet is perhaps an even deeper one: Why Mars?

In their dreamlike way, works of fiction set on Mars always raise that question by *not* raising it. It is left to the critics and historians of literature, film, and popular culture to examine and attempt to explain why the Red Planet has managed to hold such a grip on the human imagination for so long. Part of what you'll find in this volume are indeed such examinations and explanations (see, among others, Slusser on Rosny's Mars, Rabkin on Bradbury's Mars, several commentators on Frederik Pohl's version of "becoming Martian" in his *Man Plus*, Palmer on Robinson's Mars, and Robinson on his own Mars).

In what the scientific community has "seen," Mars was initially not even the best candidate for a second life-supporting world in our solar system. As noted in several essays contained in this volume, Venus — significantly closer to Earth in terms of size, mass, gravity, distance from the Sun and actual travel time — at first seemed the more likely choice. It was argued to be such until radar and radio telescopy pierced the thick Venusian atmosphere and probes finally confirmed the merciless heat of the Venusian surface.

At best, living on Mars would be distantly analogous to living in some of the most desolate places on Earth: the Atacama, the top of Everest, the Rub' al Khali (Empty Quarter), Death Valley, inside ancient meteor-impact craters in the Arctic or Antarctic — except for the fact that Mars has far less water, far less oxygen, far more violent temperature swings, and far longer supply lines than any of those (not mere tens or hundreds of miles long, but tens of millions of miles long). No single locale on Earth is hostile enough to replicate even the least hostile environment on Mars.

Yet the funhouse mirror we of Earth have most often gazed into has been that same distant Mars, both because of and despite reasons alluded to in several essays here (see particularly the discussions of the interrelationships between the scientific and science fictional visions of Mars in the Miller, LaBare, and Hendrix essays). This idea of a distant funhouse mirror is not only relevant to Mars, either. The literary and scientific traditions surrounding the Martians and their disappearance — especially those of the "Dying Planet" phase of human theorizing about Mars — bear an uncanny resemblance to similar theorizing by modern humans concerning the disappearance of the Neanderthals, in that both very clearly reflect our own contemporary cultural concerns and anxieties about the human future.

Part of our fascination with the Neanderthals, like our fascination with greenhouse-blasted Venus or with Martians of the Dying Planet scenario, arises from forebodings that such scenarios present ourselves and our world as viewed — both seen *and* dreamed — through a funhouse mirror, distorted by distance in time and or space. This aesthetic distancing allows us to bring before our minds our persistent but usually unspoken anxieties that, for all humanity's vaunted sophistication and uniqueness, we (like our presumably vanished nearest hominin relative) are not immune to extinction. In some perhaps not-so-distant future, our world might well become less the "Goldilocks" just-right planet between Venus and Mars, and more a hostile "bear" whose environment is notably less friendly toward humanity.

Despite the fact that Martians in fiction so often haunt the humans who venture into their world (especially in stories of the dying Mars), it is important to remember that the Martian funhouse mirror shows us many different images. We see in that mirror not only images which are hauntingly nostalgic (most prominently Ray Bradbury's *Martian Chronicles*) but also those which are chillingly tragic (D.G. Compton's *Farewell, Earth's Bliss*), or those which carefully rework previous histories of human ordeals in exploration (Geoffrey Landis's *Mars Crossing*) or in hyper-engineering (Greg Bear's *Moving Mars*). In the visionary funhouse, there is even room for the hilarious (including "sentient tumbleweeds" in stories ranging from Clarke's reimagining of the space program in *The Sands of Mars* to episodes of *The Outer Limits*).

Even within the "dying Mars" motif alone, it's not just "Earthlings bad, Martians good." As several authors in this volume point out, Wells's Martians in *The War of the Worlds* perform for much of that book the role of European imperialists, while the European imperial cultures are (at the hands of the Martians) forced to suffer deaths and decimations similar to those the Europeans themselves had already visited upon native peoples of the Americas, Asia, Africa, and the Pacific. The destructiveness of the Martians in Wells is a mirror stand-in for the destructiveness of many human beings in many eras.

In that other great archetype of Anglophone stories involving dying-planet Martians, their "dying race" status prompts not bellicosity but nostalgia: The experience of Bradbury's Martians in *The Martian Chronicles* (and to a lesser extent the humans who colonize their world) is clearly informed by the American frontier experience from both the Native American and European colonial perspectives.

Because we look into any mirror to find out how *we* are doing by how "they" (the images in the mirror) are doing, there is inherently an element of anxiety which must prompt that gaze — and looking into the Martian mirror is no different. We might, in specific cases involving the future of our species, not be far wrong defining anxiety as "guilt about the future." In this case "guilt" is a state of uneasiness and remorseful awareness concerned not so much with the near-certainty that we have already done something wrong — historical genocides, extinctions and near-extinctions of peoples as different as the Tasmanians, aboriginal peoples of the New World and Australasia, the Jews of Europe, and perhaps even the problematic extinctions of our nearest hominin relatives — but rather with the growing suspicion that we will have *failed* to do something *right* (questions surrounding global climate change, continued steep growth of resource consumption, and the impact of human population hitting seven billion and still rising all come to mind).

Under such a definition, the sixth great Age of Extinction (brought to all living things courtesy of modern human beings) becomes even more a cause for "future guilt." If we are not only driving other living creatures to extinction but, as a result of our activities, we someday also end up destroying ourselves — and if we are the "consciousness of Earth," as Kim Stanley Robinson contends in this volume — then the extinction of modern humans would be more than just the extinction of modern humans. It would be the extinction of all consciousness ever possessed by every hominin species and every *Homo sapiens* group with which we surviving human lines interbred or interacted (whether lovingly or genocidally). The consciousness of Earth — us — will have evolved only so that the consciousness of Earth — we — might become extinct.

Put in that "consciousness as a self-consuming artifact" way, it does all sound a bit futile. Rather than confirm the sentiment that it would have been better such consciousness had never been born, however, theories and fictions —*visions*— of the past and future which reflect contemporary cultural fears, hopes, guilts and anxieties are, at bottom, inherently gestures against such apparent futility and cynicism. By presenting, as earlier suggested, the *here and now* as reflected from *there* (a distance of space — Mars, for instance) *and then* (a distance of time, whether the past of the Neanderthals or future of the Martians), such scientific theories and scientifically informed fictions make less and less plausible our arguments that we "know not what we do." That phrase should remind us that another angle or tilt of the funhouse mirror reveals that there is also a spiritual or transcendent dimension to many of the visionary portrayals of Mars and the Martian situation (see here particularly Huntington's discussion of the uses of Mars in C.S. Lewis's *Out of the Silent Planet* and Olaf Stapledon's *Last and First Men*).

Given the proliferation of scientific explanations of present circumstances which invoke arguably similar past situations, and the "If this goes on ..." of science fiction scenarios which continue to suggest plausible (and sometimes very dark) futures following logically from our present actions and inactions, as time passes we must inevitably find it more and more difficult to plead ignorance. That we must be certain "beyond a shadow of a doubt" before we act concerning particular global outcomes or necessities increasingly looks like rationalization at best and specious excuse-making at worst. The idea that we can dump difficult problems into a global "headache box" for the next generation to open, address, decipher and defuse becomes, with each succeeding generation, more and more notoriously irresponsible. Especially in the form of the cautionary tale, science fiction is a genre which encourages responsibility by forcing us to unlock and examine the contents of our particular generation's headache box.

Recent science fictions and scientific speculations about Mars often reflect multiple items taken out of our headache box, especially in the form of the science fiction cautionary tale, a subgenre which encourages an unlocking and examination of each generation's headache box. Our headache box today is filled not just with keys missing their locks, locks missing their keys, or unidentifiable parts from forgotten or yet-to-be-assembled motors, but also with questions of what to do or what *can* be done about so many things: the proliferation of nuclear weapons and nuclear waste; habitat-destroying human population growth; accelerating extinction rates of increasing numbers of species; the collapse of ocean fisheries; growing scarcities of readily accessible energetic and material resources; unintended anthropogenic alteration of climate; perils of hard radiation to long-duration space crews and colonists of thin-atmosphered planets like Mars; the high improbability of faster-than-light travel; and all this, before we even begin thinking about big rocks from space, or gamma ray bursters.

Beyond providing grounds for interesting hypotheses, such headache-box materials — relocated to elsewhere/elsewhen places like Mars — always have the potential to become the bases for stories and developed scientific speculations which are, as it were, black-box recordings of a crash, tragedy, or catastrophe *which is yet to occur*. One purpose of such stories and speculations is in fact to prevent those horrible events from occurring in the first place. Cautionary tales succeed best when they fail — when they become the perfect antitheses of self-fulfilling prophecies, helping to prevent the very future they predicted.

Such cautionary tales would seem to be at best self-consuming and ephemeral artifacts, even as they are intended to stave off self-consumption and ephemerality at much larger scales. Curiously, however, we still value and continue to read cautionary tales like Zamyatin's *We* and Orwell's *Nineteen Eighty-Four* even when they have not proven to be "predictive" in the strict sense and have in fact been "falsified" by consensus chronological history. Given its relationship to time, the science fiction story more generally — quite beyond the subgenre of the cautionary tale — might also at first blush seem to be self-consuming and chronologically ephemeral. Most of the elements of stories peculiarly focused on the near- to mid-future are, after all, far more likely to prove "false" than "true." Any science fiction story, once thus falsified, would seem to become at best only an "alternate history" story, a subgenre whose settings are always a priori false, at least in terms of consensus historical reality.

Yet it is not necessarily the case that yesterday's cautionary tale can *only* become tomorrow's alternate history, or that even a near-future science fiction story which provides names, dates, and places can have no value beyond a certain "shelf life" or "use by" date. Precisely because visions as ephemeral as dreams can yet leave lasting impressions, "shelf life" is certainly not the whole of the matter.

Some of the stories discussed here have not yet been falsified by science or history. Some have been. Yet even those which have already been scientifically or chronologically falsified can still ring true literarily, as *stories*. The reading of Wells or Zamyatin or Orwell may increasingly require footnotes — but, then again, so does reading Dante or Spenser or Milton.

In this volume you will find a panoply of critical essays, lectures and panels. You will find literary analyses here of French Mars, Russian Mars, capitalist Mars, communist Mars, British Mars, and American Mars, to name a few. You will find the Mars of Burroughs through Brackett, of Heinlein, of Dick, and of Robinson, among many others. Most of all, you will find examinations of how these Martian-mirror stories reflect the concerns of their authors and the cultural concerns of the times and places in which those authors wrote.

The face in the mirror is always both what we see of ourselves and what we dream of ourselves. Through analyzing these dream visions — fictional experiences of Mars set in these authors' futures — we come to better understand the historical role of such fiction on Earth. The end of looking into this funhouse mirror — the tradition of Mars fictions, the tradition of scientific understandings of Mars, and this book analyzing both — is paradoxically to see more clearly: to achieve a more accurate understanding not only of a face on Mars, but also of what the world we already have (and have made) here on Earth really does and does not look like.

What makes this mirror particularly "funhouse," however, is that it is also a hammer. The statement that "Art is not a mirror held up to reality, but a hammer with which to shape it" (variously attributed to both Bertolt Brecht and Vladimir Mayakovsky) inherently recognizes that art is, in fact, both. As Kim Stanley Robinson rightly remarks elsewhere in this volume, it is "crazy and immoral" to say that Mars could serve as a quick and easy "bolt hole" or "escape hatch" (151) out of the challenges facing us here on Earth. Also elsewhere in this volume, Sha LaBare is right in asserting that "If, as William Gibson once remarked, SF does not so much predict the future as colonize it, then understanding and critiquing the worldviews we use in our colonizing is a vital action in the now, one that opens up amazing possibilities — both here on Earth, and on Mars" (157). Those quotes are all good so far as they go, but do they go far enough?

Science fiction is perhaps less a means by which the present colonizes the future than it is a means by which the future colonizes the present. As our human worldscape becomes more techno-rationalized, more techno-scientific (in itself a science fictional process), the role of the present increasingly becomes that of reverse-engineering "the" future from among those futures presented by science fiction writers, futurists, scientists and other professional temporal speculators. The future is always already "present in the present" because what the future looks like is increasingly a message which, as a result of our imaginings of it, has already been *pre-sent*.

If Einstein's contention that imagination is more important than knowledge is to make any real sense, it is in the sense that imagination is dreaming with a purpose, a dream-vision, a mirror-hammer. Only by clearly seeing our own reflections in the futures we envision can we also shape the present into something we might like to see.

Isn't this clarity what vision is ultimately all about? Such clarity as *Visions of Mars* achieves is the product of the work not only of the editors, scholars, panelists, and presenters included in this volume, but also those responsible for planning and launching the 2008 Eaton Science Fiction Conference, "Chronicling Mars," sponsored by the UCR Libraries and the College of Humanities, Arts & Social Sciences, University of California, Riverside. The conference provided the seed for the work harvested herein. Without the efforts of the institutions and individuals involved in that conference, this volume would not exist.

One

Approaching Mars

Mars of Science, Mars of Dreams
Joseph D. Miller

The Periods of Martian Exploration in Science and Literature

Mars has been a Rorschach test for the imaginative since at least the time of the Italian astronomer, Giovanni Schiaparelli. Schiaparelli, in 1877, observed what he believed to be straight lines (*canali*, or channels) in equatorial regions of Mars. Percival Lowell is often credited for extrapolating the Italian *canali* to the English canal, but actually canal is a secondary definition of *canale*. In any event other astronomers concurred with the opinions of Schiaparelli and Lowell, including Charles E. Burton, Charles A. Young and W.H. Pickering. However, the maps produced by these astronomical cartographers never agreed in detail. It is difficult to understand exactly what features these observers were "seeing," and today these shared illusions are generally thought to reflect the poor resolution of the telescopes of the time. Nonetheless it is important to understand that these events took place in a historical context in which seasonal changes in the Martian icecaps were easily observed, as were seasonal changes in broad surface features, the supposed green color of which suggested advancing and retreating vegetation. (In all likelihood these "features" were actually seasonal dust storms, although again it is not clear why multiple astronomers reported them as the color green.)

Still, a number of astronomers such as E.E. Barnard in 1903 and Eugene Antoniadi in 1909, in concert with the development of better telescopes, completely failed to see canals or green vegetation. In 1907 the famed British naturalist Alfred Russell Wallace pointed out that the low temperature, extremely thin atmosphere, and failure to find spectrographic evidence of water vapor in the Martian atmosphere probably made Mars an inhospitable habitat for life.

But in works of fiction this was still the Romantic Age of Martian literature, informed both by the fantastic visions of Schiaparelli et al. as well as the developing body of contradictory evidence. Thus, Wells could write in 1897 of a Martian civilization in *War of the Worlds*, but it was a dying civilization in a drying and dying world. Edgar Rice Burroughs in *A Princess of Mars* in 1912 has a mostly desert Mars, but the canals persist as a background for one of the most florid Martian ecologies ever described in science fiction. Perhaps this persistence indicates the continuing general popularity of the idea of a habitable Mars and a general ignorance of the weight of scientific opinion. But in all fairness the scientific evidence for a barren Mars was still mostly indirect. And so there was enough scientific "wiggle room" for the geographically restricted oases of life in C.S. Lewis' *Out of the Silent Planet* (1938), the modestly habitable desert of Stanley Weinbaum's *A Martian Odyssey* (1934), Arthur C. Clarke's *The Sands of Mars* (1951), Robert Heinlein's *Red Planet* (1949) and *Stranger in a Strange Land* (1961) or even the lush waterways of Ray Bradbury's *The Martian Chronicles* (1950). One of the last and fitting stories of the Romantic Age was Roger Zelazny's elegiac short story "A Rose for Ecclesiastes" (1963). Though it is another dying Mars story, in this lyrical piece the Martians hark back to the near-humans of Burroughs, capable as Dejah Thoris of interbreeding with terrestrial humans.

The Mars of Romance

But the Romantic Age comes to an abrupt end with the first *Mariner* mission of 1964. To the surprise of many, Mars was revealed as a cratered, moonlike planet, lacking in seas or anything looking remotely like vegetation, These results were confirmed by further *Mariner* fly-bys in 1969 and 1971 as well as the Soviet Mars fly-bys of 1971 and 1973. Thus begins the Sterile Period in the history of Mars exploration. Except for a few days in July 1976, when initial results from the *Viking* landers suggested that Mars life had been detected (an interpretation soon to be discredited), this was a period in which prevailing scientific opinion was that Mars was as dead as Luna. In fact, at one point NASA sent out a request for proposals for future Mars missions in which anything but life detection experiments was a permissible topic! The Sterile Period persisted for about thirty years, ending with the Global Surveyor mission in 1996.

In this Sterile Period Mars fiction got a lot more gritty. Novels like Lewis Shiner's *Frontera* (1984) and Jack Williamson's *Beachhead* (1992) were set against a grim, forbidding Martian background, and life on Mars took a backseat to the challenges of colonization. One of the few novels of this era to consider the possible existence of at least microbial life (perhaps stimulated by an optimistic reading of the *Viking* results) was Ian Watson's *The Martian Inca* (1977). In Greg Bear's *Moving Mars* (1993) there are Martian life-forms, but they are in a kind of hibernation until the planet is warmed by moving it close to the Sun. This idea of making Mars more hospitable to life by altering it, particularly by terraforming it, reached its ultimate flower in Kim Stanley Robinson's Mars trilogy—*Red Mars* (1992), *Green Mars* (1993), and *Blue Mars* (1996). But of course full-scale terraforming of Mars might well destroy any extant life-forms, so the conceit of the Mars trilogy is that Mars is entirely sterile. In contrast is the idea of "areoforming" or genetic and surgical modification of humans to adapt them to the Martian environment instead of the other way round. This is brilliantly explored in Frederik Pohl's *Man Plus* (1976).

For a variety of reasons to be discussed below, the period extending from about 1996 to the present is a period when again there is considerable speculation about the possibility of Mars life, at least at the microbial level. This period could well be called the Realistic Period. Mars continues to be a largely inhospitable environment, as in Greg Landis' *Mars Crossing* (2000). But now the possibility of microbial life has become fashionable in exobiology and is reflected in works such as Greg Benford's *The Martian Race* (1999) and Paul McAuley's *The Secret of Life* (2001). In the Benford novel an underground ecology, protected from the intense ultraviolet radiation of the surface, allows the proliferation of microbial forms and even an advanced form of intercellular organization, the microbial mat, the possible harbinger of multicellular lifeforms on Earth. But how have the prospects for life on Mars changed in the current era and how has that paradigm shift been expressed in science fiction? To begin to answer these questions we must first consider the prerequisites for life as we know it in any environment.

The CHOSEN Hypothesis

The only example we have of life is terrestrial. Therefore, we are largely constrained to a consideration of "earth-like" extraterrestrial life. While other possibilities exist, Mars life is usually hypothesized to be carbon-based with a similar requirement for H_2O in the liquid phase. Thus the elements carbon (C), hydrogen (H) and oxygen (O) are essential ingredients. Taken together these first three elements are necessary for making carbohydrates and fats. Nitrogen (N) is necessary for synthesis of nucleic acids, amino acids, and protein. Sulfur (S) is a necessary constituent of many enzymes and certain amino acids. Finally an energy (E) source is necessary to drive chemical reactions. On earth that energy is primarily provided by the sun, although hydrothermal vent organisms can use the energy associated with magma-heated hot springs in the deep ocean.

But there are some wrinkles in this simple picture. Phosphorus is coming to be seen as increasingly important in general metabolism. Some terrestrial organisms require vanishingly little liquid H_2O. Some bacteria are able to survive in solid ice, apparently inhabiting tiny liquid water microdomains. Other bacteria appear to thrive in solid rock, as much as two miles below the surface of the earth, at incredible temperatures and pressures. In fact, one of the reasons interest in hypothetical Mars life has revived is the discovery of abundant extremophiles on Earth. Such microbes can live in acidic environments up to a pH of 1.0, a wide range of temperatures from below freezing to the boiling point of water and in extremely salty brines. Methanogens — methane-generating organisms, found (among other places) in Utah desert soil, an environment similar in many ways to Mars — can thrive in an oxygen-free environment, using carbon dioxide as a substrate for metabolism. Perhaps most peculiar of all, extremophiles found at Mono Lake in California have now been evolved to use arsenic in their metabolic processes in place of the phosphorus found in all previously studied organisms.

The other side of the coin is natural contamination of terrestrial ecology with Mars microbes, possibly piggybacking on meteorites of Martian origin. This is energetically more favorable than terrestrial meteorites impacting Mars, because the gravitational field of Mars is much weaker than that of Earth. McKay et al. claimed to see evidence of Mars biology in a Martian meteorite in 1996, including carbonate globules, magnetite inclusions and iron sulfide, all of potential biological origin. Indeed, hematite (an iron oxide chemically similar to magnetite) globules ("blackberries"), carbonate, and sulfates have all been subsequently found in Mars soil. However, there are other explanations for McKay's observations, including terrestrial microbe contamination of the meteorite and abiological synthesis of the various minerals.

A more compelling observation in McKay's study was a wormlike segmented putative fossil with similarity to terrestrial bacteria. However, this finding was discounted since the "worm" was only about 50–100 nanometers long, thought at that time to be too small for a bacterium. But more recently, there have been observations of nanobacteria of such dimension, some of which dwell deep below Earth's surface. McKay's observations are largely discounted today; nonetheless, this work reawakened interest in the possibility of Mars life.

The CHOSEN hypothesis for Mars life is not without problems. One of the biggest is that complex organics have yet to be detected in Mars soil, although technical problems may have prevented such observation. Methane has been detected in the Martian atmosphere by three separate groups (Mumma et al., Formisano et al., Krasnopolskya, Maillard and Oewn). Since Mars appears to have no volcanic activity, the most likely generator of a constant methane signal in the upper atmosphere is biological activity. Interestingly, the areas of highest methane production overlay areas exhibiting large deposits of water ice. There is also a report of detection of formaldehyde (Peplow), another potential sign of biology, in the Martian atmosphere.

Possibly more damaging is the failure thus far to see nitrates or nitrites in Mars soil. On Earth certain bacteria fix atmospheric nitrogen into ammonia, nitrites and nitrates which more complex organisms can use for synthesis of amino acids and proteins. In the absence of nitrogen fixation no mechanism is known for getting nitrogen out of the atmosphere into a biologically available form.

Finally we must admit that although the CHOSEN hypothesis seems to determine prerequisites for terrestrial life, we have no evidence that life is inevitable in the presence of these ingredients. In the laboratory this recipe results in simple amino acids; similar constituents in the atmosphere of Saturn's moon Titan have produced fairly complex organic molecules named tholins by Carl Sagan. On Earth we know from the fossil record of stromatolites and estimates of the era in which the Earth's surface finally cooled sufficiently that the minimum time for evolution of the earliest known life is 100 million years. We don't know if that estimate applies elsewhere and if it does, whether that inevitably leads to the evolution of DNA as the primary molecule of genetic memory (in silico simulations do suggest that of all candidates so far, including such exotics as RNA ribozymes, DNA is by far the most accurate storage molecule).

The *Viking* Controversy

The *Viking* landers *I* and *II* landed on Mars on July 4, 1976. They remain the only missions to ever perform life detection experiments on another planet. The *Viking* Project was formally initiated in 1968, but planning had begun several years before. Therefore it may be argued that the *Viking* life detection experiments were the last gasp of the Romantic Age, but were interpreted according to the prevailing paradigm of the Sterile Period.

Much of the *Viking* controversy revolves around an instrument called the gas chromatograph/mass spectrometer (GCMS). Soil samples were heated to as high as 500 degrees Celsius and the resulting gases separated over time by a chromatography column. The separated gases were then ionized and further classified by their mass/charge ratio in the mass spectrometer. It was thought that any organic, carbon-containing compounds in the soil would be easily identified. In reality, the only molecules detected were water and carbon dioxide. The absence of complex organics was taken to indicate an absence of biology.

However, the GCMS was shown to be a very insensitive instrument (Navarro-Gonzalez et al.). The identical device was unable to detect organic material in an Antarctic soil sample with a known microbial population. Recently it has been suggested that iron compounds in the soil may have so efficiently catalyzed the oxidation of organic compounds that at the relatively low temperatures employed, nothing but carbon dioxide could have been detected (Navarro-Gonzalez et al., 2006). Furthermore, spectrographic data from the European Mars Express mission now indicates that 2–3 percent of Mars soil is composed of carbonates. Carbonates were not detected by the *Viking* GCMS.

The case for biology was not helped by the failure of two of the three other life detection experiments to distinguish between normal Martian soil and heat-sterilized soil samples in terms of potential biological mechanisms such as oxygen production when exposed to water vapor or carbon dioxide uptake when exposed to a nutrient solution.

But the third life detection experiment, the labeled release experiment, was very different. In this experiment soil samples were mixed with a small amount of a nutrient solution containing carbon-14 radiolabeled amino acids and carbohydrates. A scintillation counter for detecting radioactivity was positioned about one foot over the sample. Since all carbon atoms in the nutrient were radioactively labeled, the detection of radioactivity above background level by the scintillation counter would indicate the presence of a radioactive carbon–containing gas such as carbon dioxide or methane, presumably produced by cellular metabolism in the soil sample.

Indeed, when the soil sample was injected with nutrient, there was a massive release of gas which was replicated a total of four times with four different soil samples at two different sites on Mars. But if a separate sample was heat-sterilized to 160 degrees Celsius, no gas release occurred. Partial sterilization to 46 degrees Celsius attenuated but did not destroy the signal. And yet another sample stored in the dark for four months (starved?) gave the same result as full sterilization.

Yet another aspect of these data has recently been investigated (Miller, Straat and Levin). Circadian rhythms are near-24-hour oscillations in virtually every physiological and biochemical parameter studied in Earthly organisms. They are present in every terrestrial species studied, including primitive cyanobacteria. Thus, circadian rhythmicity is an excellent biosignature. Remarkably, such rhythms are superimposed on the "growth curve" of released gas in these experiments, analogously to the superimposition of such oscillations on growth curves of terrestrial cyanobacteria. Over the first couple of days of the labeled release experiments, the amplitude of the rhythm grows, suggesting the synchronization of rhythmicity in a population of microbes. Sterilization or storage in the dark destroys the Martian rhythms as well as the overall magnitude of the release signal. In ground-based experiments on Earth, terrestrial soil samples exposed to nutrient exhibit the same rhythmicity, which can again be destroyed by sterilization. Furthermore, the terrestrial rhythms are near 24 hours in period; the Martian rhythms are near 24.7 hour, where 24.7 hours is the length of a Martian day.

But a biological interpretation of the *Viking* data seemed to be disallowed by the negative results of the GCMS. Instead, it was hypothesized that Mars soil is rich in peroxides or superoxides which would oxidize any carbon-containing molecule, releasing carbon dioxide and explaining the large immediate signal in the labeled release experiment following the administration of nutrient medium to the soil sample. However, the superoxides/peroxides were entirely hypothetical constructs. To date no such superoxides have been detected in Mars soil by spectrographic analysis via the Hubble Space Telescope, Mars Express, or, apparently, the current Phoenix mission. (A less oxidizing agent, perchlorate, was detected, however; see below.) A superoxide was finally synthesized in the laboratory several years ago. However, it was completely unstable in aqueous medium and broke down in less than a second. The nutrient used in the labeled release experiment, in contrast, was indeed an aqueous solution. But the overall labeled release signal, as well as the circadian rhythm in that signal, persisted for over forty days in some of the *Viking* labeled release experiments. Occam's famous razor suggests that an insensitive GCMS is a more likely explanation of apparent failure to find organic compounds in Mars soil, coupled with a successful labeled release experiment, than a sterile Mars rich in a chemical compound that has yet to be observed!

The Phoenix Mission

The latest Mars lander was part of the Phoenix mission. Experiments conducted by Phoenix were not truly life detection experiments but rather aimed at identifying such things as soil and water ice chemistry. The Microscopy, Electrochemistry and Conductivity (MECA) instrument package allowed analysis of soil constituents, pH, and microscopic observations of soil particles. The Thermal and Evolved Gas Analyzer (TEGA) was a combination furnace and mass spectrometer, in many ways version 2 of the *Viking* GCMS. It was a much more sensitive instrument than the GCMS and able to detect organic molecules at very low concentrations.

MECA found that Mars soil is characterized by abundant atomic species such as magnesium, sodium, potassium and chlorine, largely confirming *Viking* observations (no superoxides or peroxides have been reported). These elements exist as salts in a concentration in polar soil of about 0.1 percent, at least 35 times less than the salinity of terrestrial ocean water. At lower latitudes, where the *Viking* craft landed, the soil was much more salty, with concentrations of 10–20 percent. This suggests that any water in the liquid state at those lower and warmer latitudes would be very briny. Brines will not sublimate as quickly as pure water; thus brines could persist in certain Mars regions as long as several days at low altitudes in deep canyons where atmospheric pressure will be higher. This could be a common condition in the Martian summer at equatorial latitudes. However, it is also true that most terrestrial microbes cannot tolerate very high salinity. Of course, Martian microbes may have adapted to high salinity over evolutionary history.

One surprise was the discovery of perchlorate in some soil samples. Perchlorate is a weak oxidizer. If soil samples with organic molecules and perchlorate were heated in the TEGA oven (see below) it is very possible that the oxidant may have destroyed the organics. But explaining the *Viking* labeled release studies in terms of perchlorate oxidation of organics is chemically unreasonable since strong and rhythmic evolution of gas was present for over 40 days in such studies, whereas a perchlorate effect would have been immediate. In addition, some terrestrial microbes use perchlorate as an energy source. Interestingly, perchlorate, like salt, lowers the freezing point of water. That could allow perchlorate/water mixtures to exist in the liquid state in certain locales, permitting the possibility of an aqueous biochemistry. All such speculation must be tempered by the possibility that the perchlorate was not of Martian origin, but rather a contaminant from the booster rockets which used it as a fuel.

Another surprise was that Mars polar soil is alkaline with a pH between 8 and 9. This was unexpected. Two of the *Viking* experiments suggested that Mars soil can absorb carbon dioxide, which generally occurs with acidic soils. It is possible that pH varies from region to region on Mars. It is also possible that the carbon dioxide uptake seen in *Viking* experiments may have reflected microbial metabolism. In any case, the media reported that asparagus would grow well in Mars polar soil, although this would probably also require protection from ultraviolet radiation, heating to a reasonable temperature and probably nitrate fertilizer!

TEGA was unable to detect any organic compounds. However, problems with TEGA have occurred. The first soil sample was too "cohesive" to push past the small TEGA oven hatch. However, after several days and mechanical shaking, which apparently caused a short circuit, some soil particles finally entered the oven, but only water vapor was detected with heating. It is very possible that the "cohesiveness" of the soil indicated the sample was mud! As the sample was collected by the robot arm, ice particles may have liquefied due to friction, creating a muddy suspension. Over the next day or so, as the sample dried, eventually cohesion would dissipate and particulate matter would enter the oven. Prospects for further TEGA results progressively dimmed; it is ironic that again a GCMS-like instrument underwent such technical difficulty.

The "Real" Martians

What can we say about the prospects for life on Mars? How might Mars life be realistically portrayed in modern science fiction? To answer these questions, we must first "follow the water." Channel features and gullies on Mars suggest that significant amounts of water must have flowed on the surface of Mars as recently as 100,000 years ago (some have said as late as yesterday!). The presence of carbonates, hematite ("blueberries") and sulfate in Mars soil have been interpreted as the condensation products of evaporating, possibly CO^2-laden oceans. Since Mars is about as old as Earth, this implies that water in the liquid state would have been present in some locales at least sporadically for billions of years. Since the minimum time necessary to evolve microbes on Earth is thought to be about 100 million years, there was abundant time for life to evolve in an aqueous environment on Mars. If life ever got a foothold on Mars, it is reasonable to expect that it would have evolved and adapted to progressively less surface water and a progressive increase in salinity of whatever surface water remained. On this planet we have extremophiles which can tolerate very briny environments or even the virtual absence of water altogether, as well as very low temperatures, virtual absence of oxygen, high levels of ultraviolet radiation, and so on. Of course we don't know whether life ever got that initial foothold on Mars, since we know nothing of the probabilities of making a transition from nonlife to life. At present, we have but one planetary example of that transition!

We assume in all of this terrestrial, carbon-based life, amino acids, proteins and nucleic acids similar to terrestrial species. But the metabolic function of Mars life could be as varied as that of terrestrial microbes. Thus, Martian microbes might "eat" carbon dioxide and excrete methane as in terrestrial methanogens. Or such microbes could reduce iron or metabolize sulfate, as do other terrestrial microbes. And there are abundant sulfates in Mars soil, as well as various oxidized and reduced states of iron.

But where are the nitrates, where are the complex organic molecules? It may be that we have to look in the right places. Certain regions, such as the caldera edge of the immense extinct volcano Mons Olympus, are relatively protected from ultraviolet radiation through much of the day. The same is true of the "shady" sides of vast canyons like Valles Marineris. The low altitude of these regions also suggests higher atmospheric pressure, which would allow water or brines to avoid sublimation for lengthy periods.

It is also possible that Mars microbes might be able to fix nitrogen from the atmosphere directly into protein, avoiding nitrate and nitrite production. But then we should be able to find evidence of protein, at least if we look in the right places.

It is true that methane is present in the atmosphere and probably ammonia and formaldehyde. Since these gases are broken down in the upper atmosphere by ultraviolet radiation, there must be some constant source of replenishment. In the absence of volcanic activity, biology is an excellent alternative.

So microbial life on Mars, sub-surface or in other relatively protected environments is very plausible and certainly usable as a plot device in science fiction. Can we go any further? Given the initial evolution of microbes, it is reasonable to expect evolution of colonial networks of interacting cells, what are called biofilms on Earth. It may be that Mars life has progressed even beyond that point to multicellularity and true eukaryotic life-forms. One of the simplest animals, the placozoan, is an animal with only four cell types. Mars life could have progressed this far or farther in several billion years. Many of the simplest eukaryotes, such as sponges and jellyfish, are restricted to aqueous media. However, there are iceworms on Earth which dwell inside glaciers. One can imagine iceworms in the polar ice of Mars, perhaps metabolizing CO^2 or sulfates, rather than the oxygen-based respiration of terrestrial animals. Such creatures might be slow-moving or dormant most of the time, since these more exotic forms of metabolism do not seem to be as energetic as oxygen-based respiration.

The far dispersal of species and their genes seems to allow the invasion of unoccupied ecological niches, further adaptation to those niches and subsequent speciation, leading to the vast heterogeneity of life on Earth. Such mechanisms might be at work on Mars. Many microbes can enter a sporiform stage in which they are protected against ionizing radiation and can survive without any source of energy for thousands of years. Perhaps most Mars life exists in such a state most of the time, waiting for transient increases in moisture and temperature in order to enter a rapid period of growth, reproduction and dispersal. A major means of genetic dispersion on Mars might be the sandstorms and windstorms which could disperse sporiform species widely over the planet, increasing the chances of finding at least a temporarily hospitable environment.

Might such organisms be deadly to terrestrial life? It seems unlikely considering the vast differences in evolutionary history, even if Martian and terrestrial life share the same panspermic origin. On the other hand, it is odd that anaerobic microbes can not only exist in the human body but can sometimes cause sufficient destruction of tissue (e.g., gangrene, necrotizing fasciulitis) as to result in death, all in spite of an evolutionary history that diverged from our eukaryotic progenitors many millions of years ago.

So in Martian science and science fiction the six-armed green humanoid is no longer tenable. However, we may still have to deal with inimical aliens many orders of magnitude smaller — a nice twist on Wells' *War of the Worlds*. Just as likely, such aliens might teach us the greatest lessons in biology since the discovery of DNA, particularly if such life-forms have a biology that is entirely different from anything we may expect on the basis of our terrestrial experience.

Mars Science and Mars Fiction

A defining characteristic of good science fiction is that the science must be plausible. It need not be mainstream "accepted wisdom," and in fact controversial science is a good spice in the fictional recipe. However, scientific plausibility is always a moving target. When all that was known of Mars were seasonal changes in the icecaps and certain other topographical features and the controversial inference of canals by Lowell and others, the door was wide open for the most florid imaginings of Wells, Burroughs, Lewis and others. But even in this Romantic Age of Mars fiction, Mars is an arid and often dying world. Nevertheless, there is no lack of native inhabitants. This is clear in Weinbaum's *Martian Odyssey*, as well as in the iconic stories of the time by Heinlein, Clarke, and Bradbury. But in 1964 *Mariner* ends the Romantic Age, a clear example of how science informs science fiction.

The Sterile Period eschews the possibility of macro-life on Mars for the most part. If it does exist it is cryptic and hibernating, as in Bear's *Moving Mars*. Microbiology is still allowable, as in Watson's novel *The Martian Inca*, but for the most part this era focuses on human colonization of an inhospitable and biologically inimical planet, as in Williamson's *Beachhead* and Shiner's *Frontera*. In Pohl's hands in *Man Plus*, the human colonist is converted to a life-form capable of a Martian existence, but indigenous forms are absent. The Sterile Period reaches its nadir with Robinson's famous Mars trilogy. Biology is absent and any objection to terraforming Mars is in terms of damage to the environment, not to indigenous life-forms. It is not a matter of chance that the period of publication of Robinson's trilogy, 1992–1996, is the low point in NASA's interest in potential life detection experiments on Mars. Again science fiction responds to the scientific zeitgeist.

The pendulum moves again to the Realistic Period. McKay's pivotal analysis of the Martian meteorite in 1996 suddenly amplified the possibility of microbial life on Mars, even though the biological interpretation of those results has largely been dismissed. However, the basic observations of carbonates, magnetite, and sulfides have all been more or less confirmed by the *Phoenix* results. Furthermore, the characterization of terrestrial extremophiles has reified the notion that biology could survive the extreme environment of Mars. And the now certain presence of water ice and perhaps liquid water in certain regions at certain times of the Martian year, as well as the detection of gases of probable biological origin in the Martian atmosphere and continuing re-interpretation of the *Viking* results, has led to a kind of renaissance in thinking about Martian exobiology. (Some argue this is more characteristic of the European Space Administration than NASA.) In science fiction, microbes are back, as in Benford's *The Martian*

Race and McAuley's *The Secret of Life*. The Benford novel suggests that Mars life may have evolved beyond the microbial stage to a kind of colonial organism, something like a terrestrial microbial mat. This is a completely legitimate scientific speculation given that if life ever evolved on Mars, it has had billions of years to progress beyond the microbial stage. Life is tough and microbial life even more so — the most inclusive terrestrial extinction event of all time, the Permian extinction, obliterated 70 percent of land organisms and perhaps 95 percent of marine organisms. But microbes were virtually untouched.

Science fiction is probably the favorite literature of the world's space scientists. Is it possible that biological speculation in science fiction could feed back into the planning of life detection experiments? NASA plans a Mars Science Lab originally scheduled for launch in 2011 and the European Space Agency plans to launch Exo Mars between 2016 and 2018. Both missions plan to look for biosignatures via gas chromatography and mass spectrometry, many generations removed from the ancestral *Viking* instruments, although neither mission includes even a simple attempt to culture microbes from soil samples or visualize them with a high-powered microscope. It appears that the membrane between the two cultures is largely one way — science informs the fiction, but even the best extrapolations from the fiction do not seem to change the mindset of those who plan the missions.

It is safe to conclude that science has narrowed the scope of what is plausible in science fiction. But our judgment of what is "good" science fiction must be historically informed. Verne's *Journey to the Center of the Earth* was plausible in his era, but constructing a vehicle out of "unobtanium" capable of withstanding a journey to the incredibly hot, dense core of the Earth, as in the 2003 movie *The Core*, is simply ridiculous.

A narrow scope provides many opportunities for plausible fiction. The most radical approach considers a more hospitable Mars in a parallel universe, as in Niven's *Rainbow Mars*. Less extreme notions focus on evolved microbial life as in Benford's novel. There is no particular reason to stop at that level; billions of years of evolution may have produced organisms like our terrestrial iceworms, capable of living in solid ice. If subsurface aqueous pools exist, evolution may have progressed far beyond microbes. And even surface life of small dimension cannot be excluded; the resolution of the Mars Orbital Camera is but 1.5 meters and the cameras on board the rovers have limited fields of view. But of course any surface life must be adapted to high ultraviolet radiation, lack of oxygen, limited water, and extreme temperatures. Something like a methanogenic cockroach could fit the bill. Even in the absence of extant life there may well be fossil evidence of Mars life, perhaps in regions of the rock strata walling the largest canyon on Mars, Valles Marineris. Martian paleontological fiction is yet to be written.

And so the free verse of the Romantic Period, Burroughs' six-armed Martians, is displaced by the tightly constrained Realistic Period haiku of Benford's microbial mats. But constraints can nevertheless allow and even stimulate aesthetic creation. In Benford's famous analogy, the net in the tennis game of Mars science fiction gets ever higher as our knowledge increases. But the game remains compelling. And in the worst case of a total absence of life on Mars we can at least say, "We will always have Europa!"

Works Cited

Alan Burdick, "Seeding the Universe." *Discover Magazine*. October 1, 2004. http://discovermagazine.com/2004/oct/seeding-the-universe/article.
David S. McKay, et al. "Search for Past Life on Mars: Possible Relic Biogenic Activity in Martian Meteorite ALH84001." Science (August 16, 1996): 924–930.
Michael Mumma, et al. "A Sensitive Search for Methane on Mars." *Bulletin of the American Astronomical Society* 35 (2003): 937.
Vittorio Formisano, et al. "Detection of Methane in the Atmosphere of Mars." Science 3 (December 2004): 1758–1761.
Vladimir A. Krasnopolskya, Jean Pierre Maillard, and Tobias C. Owen. "Detection of Methane in the Martian Atmosphere: Evidence for Life?" Icarus (December 2004): 537–547.
Mark Peplow, "Formaldehyde Claim Inflames Martian Debate." Nature. (February 25, 2005). http://www.nature.com/news/2005/050221/full/news050221-15.html.
Rafael Navarro-González, et al. "The Limitations on Organic Detection in Mars-like Soils by Thermal Volatilization–Gas Chromatography–MS and Their Implications for the Viking Results." PNAS (October 31, 2006): 16089–16094.
Joseph D. Miller, Patricia A. Straat, and Gilbert V. Levin. "Periodic Analysis of the *Viking* Lander Labeled Release Experiment." *Proceedings of SPIE Conference*, 2001. http://mars.spherix.com/spie2/Miller-Straat-Levin_FINAL.htm.

Where Is Verne's Mars?
Terry Harpold

Verne's Mars

The Red Planet is seldom discussed in the sixty-two novels and two collections of short stories that make up Jules Verne's *Extraordinary Voyages* (*Voyages extraordinaires*, 1863–1919).[1] Mars's appearances in the *Voyages* are few enough that they are easily enumerated.[2] In *From the Earth to the Moon* (*De la Terre à la Lune*, 1865) and its sequel *Around the Moon* (*Autour de la lune*, 1870), Mars is among the planets mentioned by the occupants of the space-bullet *Columbiad* as they debate several astronomical questions.[3] In the two novels, Mars is mentioned a total of seven times.

In *Face the Flag* (*Face au drapeau*, 1896), the Fulgurator, a new armament invented by megalomaniac scientist Thomas Roch, is said to be so powerful that a few thousand tons of it would be enough to destroy the Earth and send its fragments careening into space "like the exploded planet [that once orbited] between Mars and Jupiter" (ch. xii).[4] In *Five Weeks in a Balloon* (*Cinq semaines en ballon*, 1863), Joe Wilson, the novel's comic character, is enchanted with the delights of lighter-than-air travel. He speculates half in jest that it might be possible to visit Mars ("where military types are the big shots") and other planets in a balloon (ch. ix).

Robur the Conqueror (*Robur-le-Conquérant*, 1886) opens with an extended nighttime prank, as the eponymous aviator flies over major cities of the world playing a trumpet and furtively planting his banner atop their tallest structures. The cities' inhabitants are understandably disquieted, observes the narrator:

> If you heard some strange and inexplicable sounds in your house, wouldn't you try to figure out the cause? And if your inquiry was unsuccessful, wouldn't you leave your house and go live somewhere else? Of course! But in this case, the house was the terrestrial globe. There was no way to leave it for the Moon, Mars, Venus, Jupiter, or any of the other planets in the solar system [ch. i].

It was necessary, the narrator says, to look for answers to the mystery of the prank not in infinite space, but in the Earth's skies.

Verne's only interplanetary adventure, *Hector Servadac* (1877), contains the greatest number of references to the planet Mars in a single novel: fourteen.[5] But Mars is never more than a signpost on the two-year journey of the comet Gallia as it passes through the solar system. The castaways from Earth who have been swept away on the comet are too busy surveying its features and finding other survivors to pay much attention to Mars — Gallia has, in fact, moved beyond the planet's orbit before they become aware that they are no longer on the Earth's surface.[6]

Topsy-Turvy (*Sans dessus dessous*, 1889) includes an unremarkable mention of Mars in a list of the planets making up the solar system (ch. viii), and two striking mentions in one paragraph, in a unique Vernian reference to features of Martian topography. The novel's narrator compares the effects of the Baltimore Gun-Club's literally Earth-shaking plan to rotate our planet's axis by 23 degrees (the Gun-Club seeks easier access to coal reserves at the North Pole) to the inundation of the Martian Northern Hemisphere when seas shifted from its Southern Hemisphere. If, he asks, a few charitable souls on Earth were worried about drowned Martians and started a subscription for their relief, what would be their reaction to a greater catastrophe closer to home? (ch. x).[7]

In the short story "A Day in the Life of an American Journalist in the Year 2890" ("La Journée d'un journaliste américain en 2890," 1910), Mars is the source and destination of "phototélégrammes," apparently correspondence with friendly Martians, to and from the global newspaper conglomerate the *Earth-Herald*. But this minor text was written mostly by Verne's son, Michel; the senior Verne probably had not much to do with it (Gondolo della Riva 1974).

Finally, there is to my knowledge only one mention of Mars in Verne's published nonfiction. In a 1902 interview with *The London Daily Mail*, Verne is asked his opinion of the works of H.G. Wells. He hasn't read Wells, Verne replies, because he's waiting for some good French translations. He then repeats an idea he proposed earlier in the interview, that the popular novel will one day be supplanted by journalism, and adds:

> The writer of the future, even if he peoples Mars, the Moon, and other planets, can only do so [based] upon human data. The Moon men will be human beings in carnival attire, that is all. One day, perhaps, communication with the other planets will be possible, and then you will have, not novels about Mars and the moon, but your Lunar and your Martian newspaper correspondents.

"Their work," he concludes, "will be more interesting reading than either Mr. Wells's or my own" (Verne 1902).

If these few, brief references to Mars — no more than two dozen in all — tell us one thing about Verne and Mars, it is that he wasn't much interested in the place. Yet, plainly, authors writing in the traditions of imaginative fiction that he helped to initiate wrote extensively of Mars, Martian landscapes, and Martians, human and alien. Why didn't Verne? Should we be surprised that the most influential author of the scientific romance mostly ignored spaces that have become exemplary terrains of modern science fiction? What is the value of Verne's negative example of the Martian imaginary?

The most useful answers to these questions, I think, depend on fundamental patterns of Verne's fiction. With the exceptions of the two lunar novels and *Hector Servadac*, all of the *Voyages* unroll on, over, and under surfaces; even those exceptions follow clearly defined circuits that resemble at least in shape those of the terrestrial adventures. All of those nifty flying and submarine devices, the sailing ships, trains, automobiles, wagons, all that walking — and Verne's characters walk as much or more than they ride — describes *paths*, not places. Narrative action in the novels is tightly, we might say obsessively, bound to systems of motion. These may be halting, digressive, or recursive — but things in Verne's fictional universe *are always moving*.

In this respect, the shapes of Verne's adventures are, as the French philosopher and critic Michel Serres has observed, "loxodromic" (Serres 1968, 1990). Whatever local errancies they may include, their global paths return inevitably to a constant bearing; all the hazards and misfortunes faced by their heroes are strictly determined, moving from the first to the last step with a goal in mind. Which is another way of saying that they are *fictions*. For Verne's apparent realism is his canniest conceit: the adventure takes the form of a series of accidents, but a providential grace behind it all insures that things always work out, and always according to the rules of a game whose winner is determined from the start (Harpold, "Providential Grace").

This basic orientation of Verne's literary program is the key to appreciating his negative Martian example, in that in the *Voyages*, narrative form is above all a system of trajectories. Not change: very little actually changes in Verne — for example, he has almost no interest in characters' psychological evolution. Instead of changing, his actors and actions turn in carefully calibrated circuits, like the clockwork rotations of an orrery, one of those wonderfully complex contraptions that model the positions and motions of the planets. The spaces of the Vernian imaginary are determined by just that sort of turning and repeating.[8]

If Verne rarely included Mars in his fictional systems, it may be that he didn't need to go beyond terrestrial circuits in order to demonstrate the effects of his literary mechanisms. The artistic economy of staying closer to home in this way is easy to see once you accept the principle that in Verne the North and South Poles, equatorial Africa, Magellania — or for that matter, the streets of Paris, London, and Chicago — are no less literary constructs than are the dry seabeds, canals, and abandoned cities of Barsoom.[9]

Reading Mars with Verne

But I am determined to get us to Mars and back with Verne. As a small push in that direction, I suggest that Verne's methods can model for us a productive way of reading modern fictions of Mars.

You have to travel at least 35 million miles to get there, or the Martians, such as they may be, will have to travel the same distance to get here. And once you or they arrive, neither we nor they can very well stay in one place, if the aim of the trip is to be exploration or colonization. Assuming we go there, the navigable surface of Mars is equal to the land surface of Earth, so there's a lot of territory to cover when you start moving. But we shouldn't confuse such practical concerns of movement or the technologies that enable it with the imaginative or literary significance of these things. Isn't the enduring charm of NASA's rovers *Spirit* and *Opportunity* that they have kept on *moving* over the Martian surface longer than we expected them to? Doesn't their wobbly, halting parade preserve for us a fantasy of a grand tour of the planet, at least until our science catches up with our fiction?

A Vernian way of reading Martian fiction would mean thinking about the displacements of humans or aliens or vehicles as more than matters of getting from here to there or back, and addressing them in a general way as the preconditions of narrative and mimetic programs. In other words, it would mean thinking not about Martian *places*— the Red Planet as our privileged shorthand for the most basic of other scenes — but about Martian *paths* and *trajectories*, about the crossings of thresholds and the tracings of circuits, above, on, and below surfaces that happen to be Martian. And for that reason, they matter more because they happen not to be *here*.

I will end with a few examples in this regard. They're not comprehensive and I'm sure they neglect much that is interesting about the fictions they describe. But I propose that they, or other examples like them, may plot shapes of a Martian route that we could follow productively in Verne and beyond.

In Edgar Rice Burroughs's novels, John Carter travels to Barsoom by means of "astral projection," a way of moving the mind without moving the body. His adventures on Mars join the trajectory of knight errancy to the collapsing borders of empires in crisis and a planet in decline.

Ray Bradbury's *Martian Chronicles* are many things to devotees of Martian fiction, but they are at least allegories of transit, in which passage from one place to another — from Earth to Mars, through the ruins of the Martian cities, back to Earth after the outbreak of war, back to Mars and the remnant Earth settlements in war's aftermath — signals a change in consciousness that is tightly bound to spatial grammars of the modern frontier narrative.

With the invention of teleportation (the "jaunte"), Alfred Bester's *The Stars My Destination* presents us with a universe in which the projection of body and mind constitutes the destiny of 25th-century humanity. Gully Foyle's mastery of the space-jaunte not only separates space from time, but also folds narrative ends of the novel back into their preconditions. In contrast to the dizzying freedoms of the jaunte, the horror and pathos of the Skoptskies, the novel's Martian ascetics, are rooted in their total *immobility*: their sensory nervous system surgically severed from their consciousness, they live without sight, sound, speech, smell, taste, or touch.

Overall, Kim Stanley Robinson's Mars trilogy is not very Vernian; its SF is harder, and its emphases on ecological and psychological transformation have few correlates in the *Extraordinary Voyages*. But the basic procedures of Robinson's account of terraforming and the political-cultural conflicts it generates bind the surface of the planet to a general fantasy of unrestricted mobility. Areology is in this sense a sister art of a terrestrial cartography that Verne would have embraced, retracing the Martian landscape with new contours and borders defined by human agency.[10]

The Martians of H.G. Wells's *The War of the Worlds* fall to Earth and wreak havoc on humanity by moving over its cities with merciless directness. Their engines of war are dreadful expressly because their perverse, inhuman form of locomotion is unassailable, until the Martians are felled by humanity's microscopic allies, "the putrefactive and disease bacteria against which [the Martians'] systems were unprepared." We are saved because the smallest moving thing stops the strangest moving thing.

The massive corpus of UFO contactee literature — another kind of loxodromic science fiction — is run through with meetings with Martians and voyages to and from Mars. At least as far back as the U.S. 1896–97 airship "flap" — so, roughly a year before Wells's 1898 novel — such accounts associate unexpected flying craft with Martian explorations or invasions of Earth.[11] Contact narratives of the classic Adamski or Aetherius type more prominently feature Venus, but Mars is a required port of call on tours of the solar system with the Space Brothers.[12]

I'll conclude with one more example, to throw a film into this mix — no doubt there are many others that would fit here — but mostly because this one brilliantly demonstrates a program of unforgiving, impulsive movement that I think is one of Verne's most important legacies to subsequent imaginative fiction, even if the film's tenor is much darker and more disturbing than anything in Verne's oeuvre.

Roy Ward Baker's 1967 film adaptation of Nigel Kneale's *Quatermass and the Pit* resurrects our Martian ancestors as crowding, fourmillating specters: monstrous three-legged locusts, *they jump*, and the population of the modern city of London jumps in kind, in rapt lockstep, massacring "mutant" humans who are immune to the call.13 This time, the Martian devils are defeated not by humble microbes but by driving their energies to ground — literally: closing an electrical circuit that frees their human descendants from the enthrallment of a mobile destiny. Recall the film's brilliant end title sequence. After the Martians have been (temporarily?) defeated, Quatermass (Andrew Keir) and Barbara (Barbara Shelley) stand and sit, respectively, under an archway, while the ruined city burns behind them. Sirens wail in the distance. Unmoving, exhausted by their participation in the crazed melée induced by energies of the Martian spacecraft, they stare passively off-camera as the credits roll over them. They seem to be waiting for something to jolt them back into action.

Notes

1. The last ten of the *Voyages* were published after Verne's death in 1905 under the editorial supervision of his son Michel, who substantially revised or rewrote most of the posthumous works without acknowledging these interventions. For purposes of clarity, I assume in this essay that the illustrated octavo editions ("grands in-8°") of the novels published by Pierre-Jules and Louis-Jules Hetzel (64 titles, 47 double volumes), and later reprints based upon the Hetzel editions, represent the *Voyages'* canonical form. Since the late 1980s, new French editions of the posthumous works based on Jules Verne's uncorrected manuscripts have been published; several have been translated into English. These new "original versions [*versions originales*]" complicate the definition of Verne's œuvre, but they do not change the number of his discussions of the planet Mars.

2. *Mars* is also the French name of the third month of the Gregorian calendar. The more than

500 references to the month of March in the *Voyages* are excluded from the tally I give here. Indirect references to the Roman god Mars (as distinct from the planet named after him), are similarly excluded, for example: numerous references to Paris's Champ de *Mars*, the parade grounds of France's École Militaire and after 1886 the site of the Eiffel Tower; one of the characters of *North Against South* (*Nord contre Sud*, 1887) is named Mars, after the god, not the month.

3. *From the Earth to the Moon* (three times), ch. v (twice), ch. xix ; *Around the Moon* (four times), chs. v, viii, ix, xxx.

4. All translations from French are mine.

5. Part I, chs. xii, xv (twice), xvi, xxiv; Part II, chs. iii, iv (twice), viii, ix (three times), xvi, xvii.

6. In contrast, Jupiter and Saturn are given a good deal of attention in the novel because of the effects of their gravitational pull on the shape of Gallia's path.

7. We can pinpoint Verne's sources for this anecdote with fair precision. In 1888, Henri Perrotin announced his observations of the recent flooding of a "continent," which Schiaparelli had named "Libya," by the waters of the adjacent Syrtis Major, a region between the Martian southern highlands and northern lowlands. Perrotin's claim was echoed by Schiaparelli in an 1888 article (published in German, translated into French in 1889) in which he noted the diminishment of Libya during the 1880s. Camille Flammarion popularized the idea of a catastrophic Martian flood in 1888 articles for *Le Figaro* and *Astronomie*, and it appears to have been taken up more widely in the American, British, and French popular presses. (See Crowe 1999, 492–494.) An obsessive daily reader of newspapers and scientific journals, Verne would certainly have known of the flood story.

8. Harpold, "Verne's Cartographies." Again, *Servadac* is the exception that seems to prove a rule. See Weissenberg's discussion of the celebrated *cartonnage du monde solaire* ("solar system binding") of the novel's first in-8° edition (1877), which strongly resembles an orrery.

9. What were the Schiaparellian-Lowellian fantasies of "kinematic" Martian canals (Lowell 1906), and all the fictional landscapes they inspired (Serviss's "Lake of the Sun," Burroughs's Barsoom, etc.), if not paths of movement (*canali* = "channels") that permitted imaginative parcelings of the planet's surfaces?

10. Harpold, "Verne's Cartographies." See, for example, in *Red Mars* the caravans joined by Frank Chalmers in his pre–Revolution travels: "The Arabs who live out of Arabia are called Mahjaris, and the Arabs who came to Mars, the Qahiran Mahjaris.... These wanderers were mostly Bedouin Arabs, and they traveled in caravans, in a deliberate recreation of a life that had disappeared on Earth. People who lived in cities all their lives went to Mars and moved around in rovers and tents. The excuses for their ceaseless travel included the hunt for metals, areology, and trade, but it seemed clear that the important thing was the travel, the life itself" (405).

11. Cohen 1981, Busby 2004, Danalek 2010. Busby's attempt to map the paths of every airship sighted during the flap seems the epitome of SF loxodromy.

12. See Saliba 1999, Bennett 2001, Hollings 2008.

13. Rudolph Cartier's equally brilliant 1958–59 serial adaptation for BBC Television drives historical lessons of the story home by cutting in newsreel footage during these scenes, of London burning during the Blitz.

Works Cited

Baker, Roy Ward, dir. *Quatermass and the Pit [Five Million Years to Earth]*. Screenplay by Nigel Kneale. Hammer/Warner Brothers, 1967.

Bennett, Colin. *Looking for Orthon: The Story of George Adamski, The First Flying Saucer Contactee, and How He Changed the World*. New York: Paraview Press, 2001.

Bester, Alfred. *The Stars My Destination*. Eds. Alexander Eisenstein and Phyllis Eisenstein. New York: Vintage Books, 1996.
Bradbury, Ray. *The Martian Chronicles*. New York: Bantam, 1979.
Burroughs, Edgar Rice. *Under the Moons of Mars (A Princess of Mars / The Gods of Mars / The Warlord of Mars)*. Eds. Scott Beachler and James P. Hogan. Lincoln: University of Nebraska Press, 2003.
Busby, Michael. *Solving the 1897 Airship Mystery*. Gretna, LA: Pelican, 2004.
Cartier, Rudolph, dir. *Quatermass and the Pit*. Screenplay by Nigel Kneale. BBC Television, December 22, 1958–January 26, 1959.
Cohen, Daniel. *The Great Airship Mystery: A UFO of the 1890s*. New York: Dodd, Mead, 1981.
Crowe, Michael J. *The Extraterrestrial Life Debate, 1750–1900*. New York: Dover Publications, 1999.
Danalek, J. Allan. *The Great Airship of 1897: A Provocative Look at the Most Mysterious Event in Aviation History*. Kempton, IL: Adventures Unlimited Press, 2010.
Gondolo della Riva, Piero. "A propos d'une nouvelle." *L'Herne* 25 (1974).
Harpold, Terry. "The Providential Grace of Verne's *Le Testament d'un excentrique*." *IRIS* 28 (2005).
_____. "Verne's Cartographies." *Science Fiction Studies* 32.1 (2005).
Hollings, Ken. *Welcome to Mars: Fantasies of Science in the American Century, 1947–1959*. London: Strange Attractor Press, 2008.
Lowell, Percival. *Mars and Its Canals*. London: Macmillan, 1906.
Robinson, Kim Stanley. *Blue Mars*. New York: Bantam, 1996.
_____. *Green Mars*. New York: Bantam, 1995.
_____. *Red Mars*. New York: Bantam, 1993.
Saliba, John A. "The Earth Is a Dangerous Place—The World View of the Aetherius Society." *Marburg Journal of Religion* 4.2 (1999). http://www.uni-marburg.de/fb03/ivk/mjr/pdfs/1999/articles/saliba1999.pdf.
Serres, Michel. "Géodésiques de la terre et du ciel." *L'Arc* 29 (1990).
_____. "Loxodromies des *Voyages extraordinaires*." *Hermès I: La Communication*. Paris: Les Editions de Minuit, 1968.
Serviss, Garrett Putnam. "Edison's Conquest of Mars." *New York Journal* (January 12–February 10, 1898).
Verne, Jules. *Les Œuvres de Jules Verne*. 50 vols. Lausanne: Editions Rencontre, 1966–71.
Verne, Jules, and John N. Raphael. "An End to Novels," *The [London] Daily Mail*, July 5, 1902.
Weissenberg, Éric. "Le Cartonnage du monde solaire." *Bulletin de la Société Jules Verne (NS)* 35.138 (2001).
Wells, H.G. *The War of the Worlds*. Eds. Patrick Parrinder, Brian Aldiss, and Andy Sawyer. New York: Penguin, 2005.

Rosny's Mars

George Slusser

In the wake of astronomical discoveries in the last half of the 19th century, our neighbor Mars became a blank slate for speculation. From roughly the 1860s to the late 1950s, before scientific evidence reversed the belief in significant life on Mars, the focus of this speculation was the nature and variety of Martian life-forms. France, during this period, did not produce a work as resoundingly authoritative as Wells's *The War of the Worlds*. But there were a large number of Mars stories in which an important part of the adventure was the amazing flora and fauna of the Red Planet. Among these works, one stands out as radically different: J.H. Rosny's *Les Navigateurs de l'Infini* (1925). Remarkable for its time, Rosny's novel constructs a complex, minutely detailed Martian ecology, one that incorporates the fin-de-siècle vision of his earlier *La Mort de la Terre* (1910), yet at the same time looks beyond the evolutionary fatality of that work, toward a vision of humanity as interstellar force that foreshadows the later work of Heinlein and the writers of the American "Golden Age." Rosny's short novel, way ahead of its time and cultural context, reads like a 1950s space exploration. Yet one cannot call it "seminal" in any strict sense. It has not been translated into English to date, and was certainly unknown to the American writers who shaped this "classic" vision of SF mankind.

Let us first place Rosny's novel in its French context. An interesting online bibliography, *Littérature française martienne de 1865 à 1958, ou le merveilleux scientifique à l'assaut de la planète rouge* (gotomars.free.fr), visualizes dozens upon dozens of Mars novels that rolled off the popular presses during the life-on-Mars period from Lowell to *Mariner IV*. These novels span the following period: in 1865, we have an early novel from Verne's publisher Hetzel, *Un Habitant de la planète Mars*, by a certain François Peudefer de Parville; this is followed by the astronomer Camille Flammarion's "La planète Mars," in *Rêves étoilés* (1888), in which he discusses Mars as place of reincarnation; there is Arnould Galopin's *Le Docteur Oméga, aventures fantastiques de trois français dans la planète Mars* (1906), which recounts the exploits of the mad scientist of the title, the inventor of "repulsite," a form of "anti-mass propulsion"; Gustave Le

Rouge's *Le Prisonnier de la planète Mars* (1908), with sequel *La Guerre des vampires* (1909), which features an American inventor hero, Wellsian vampires, and a Burroughs-like space opera several years before *A Princess of Mars* (1912); Jean de la Hire's *Les Conquérants de Mars* (1911), which features mad scientists, Hertzian waves and utopian colonies on Mars, all in one package; Théo Varlet's *L'Épopée martienne* (1922), which replays (or rather rips off) Wells and gives us more Martian vampires; finally, in the same year as Rosny's novel, we have Henry de Graffigny's *Voyage de cinq américains dans les planètes* (1925), a space opera adventure in the "manner" of Jules Verne (Americans in space), with a lavish dose of fantastical Martian creatures. All of these works are accompanied by creatively pulpish cover art, and form the underground workshop that gave rise, in the 1950s, to the magnificent magazine covers and *bandes dessinées* of artists like Jean-Claude Forest and Philippe Druillet. What they offer however, in print and visually, is an implausible Martian environment and fantastical life-forms. Among these, only Flammarion gave the reader detailed descriptions, mixed with heavy doses of interplanetary spiritism, of Martian flora and fauna that to some degree considers environmental conditions such as scientists of the time imagined them. In contrast to these, Rosny's novel stands out as unique. For what he offers the reader is not just details, but what we would call today a complex, carefully worked out, Martian ecology. Not only are Rosny's Martian life-forms plausibly adapted to what was known of the planet's physical conditions, these life-forms are depicted as forms in evolution. Rosny's Martians develop in a context of temporal transformations, as biological entities interacting dynamically with complex changes in their environment.

But if Rosny's work is so different from that of his French contemporaries, who was he? J.H. Rosny *aîné* (Rosny senior) is a Francophone Belgian writer (1853–1940), whose real name is Joseph-Henri Boëx. Forced to interrupt his formal education, Rosny emigrated to England in the early 1880s, where he worked for seven years as a night operator for British Telegraph. Rosny apparently spent his days attending lectures and studying Darwinian evolutionary theory. His English scientific "education" seems to have placed Rosny at odds with the then prevalent Cartesian-Comtean tradition in French science. Recognizing that the English language was a barrier to his writing career, Rosny relocated to Paris in 1887. Throughout his long life, he published prolifically. His work alternated between novels in the "naturalist" vein, and speculative fictions, which range from evolutionary prehistoric extrapolations (*Les Xipéhuz*, 1887), to contemporary fictions dealing with physical mutations and perception of alternate "worlds" ("Un autre monde," 1896), to his stunning "last man" novel, *La Mort de la Terre* (1910), which chronicles the ascent and descent of carbon life on Earth, and the rise of a new, iron-based form of life, the "ferromagnetics." *Les Navigateurs de l'Infini* is Rosny's last work of speculative fiction. It is also his first work not to take place on Earth.

This novel in fact did not have an "SF" publication until 1960, when it appeared in Hachette's *Rayon fantastique* #69, edited by G.H. Gallet. Attached to this edition was a sequel, *Les Astronautes*, apparently written also in 1926 but never before published. In his introduction to *The Best of Stanley G. Weinbaum*, published in 1974, Isaac Asimov gave his overall assessment of English-language depictions of Martian life-forms up to that time. For Asimov, no work published before Weinbaum's *A Martian Odyssey* (1934) surpassed it in rendering life on Mars: "Weinbaum's easy style and his realistic description of extraterrestrial scenes and life-forms were better than anything yet seen" (Asimov ix). Even so, Weinbaum's novella is less the account of a scientific expedition than a shipwreck narrative. We are in the 21st century. Following a successful moon landing, the spaceship Ares takes off for Mars, with a four-man crew, to study the planet. The ship crashes, and the ship's chemist Dick Jarvis saves a Martian, the birdlike Tweel. On their trek across the planet, these two encounter various life-forms: silicon-based creatures that excrete bricks in order to make pyramids around them; "dream beasts" that read Jarvis's desires and project siren-like apparitions (an idea Bradbury will borrow); four-armed barrel creatures, with eyes around their waists, that push carts of stones to feed a giant "wheel." Behind the wheel, Jarvis discovers a crystalline life-form that emits a radiation that cures disease. Jarvis steals the crystal, is attacked by the barrels, and finally escapes Mars, leaving Tweel behind. The "plot" speaks for itself. The much-touted Martians, however, are a mishmash of hybrid oddities, forms that (except, perhaps, for Tweel, who flies) bear no coherent relation to a Martian environment, let alone to an ecological system in interactive evolution. To find such adaptability, Asimov would have had to look beyond Mars, to a work like Hal Clement's *Mission of Gravity*, serialized in *Astounding* in 1953.

Rosny's navigateurs (which he is the first to call *astronautes*, effectively inventing the term for future use) have a similar situation to Weinbaum's explorers. As in Weinbaum, there is considerable technological vagueness about their means of getting to Mars (their ship, the *Stellarium*, is made of "sublimated argine," said to be a transparent and indestructible material, with no further specifications). There is no mention of how the ship is propelled. It is specified, however, that there are supplies for nine months: three out, three back, three on the planet. There are three astronauts on board. The way they are delineated reminds the reader of Jules Verne's *De la terre à la lune*: There is Antoine the pure mathematician, Jean the "experimenter," and the narrator, Jacques, a "dreamer" in the manner of Michel Ardan. We learn offhand that there has been a previous moon landing. We learn obliquely, as Jacques describes the Morse code system *du dernier siècle* they use to communicate with Earth, that the time is late 20th century.

But here all similarity ends between Rosny and Weinbaum. Rosny's astronauts are scientists, not plunderers. On a real scientific mission, they approach Mars, enter the planet's gravitational field at the equator, and circle it. What they observe first is geography and climate. If their initial hypothesis is "Évolution de Mars doit ... ressembler à la nôtre [the evolution of Mars should ... resemble our own]," they prove adept at describing the strangeness of what they see. Their first impression is of a planet where water was once abundant, but which now is "plus stérile que nos déserts [more sterile than our deserts]" (Rosny 44).* Some zones, however, seem to have primitive vegetation, or "pseudo-végétations." They observe what they call "structures," chaotic amalgams

---

*All further references are to this text.

of spirals, "straps," knots, wavy lines. These seem to move, but from the air the observers cannot tell if these are forms of life. For where there should be bodies, there are "contours si irréguliers que ces êtres nous parurent informes [contours so irregulars that these beings seem to be without form]" (45). Where there should be a head, they find something that looks like "sponges," "foam."

They approach the surface and measure the Martian atmosphere. Their hygrometer detects water vapor, and a Martian "air" that is $2/7$ oxygen, $1/3$ nitrogen, with $1/10,000$ part of carbon dioxide. They conclude that there is enough oxygen to allow them to move about with the aid of their "condensers." The physical facts may be incorrect, but the general description is of a cold, dry, sterile planet. At this time, they are "frappés par un phénomène extraordinaire [struck by an extraordinary phenomenon]," a series of "phosphorescent, luminous columns" moving about their ship. They question whether these forms are alive: "Vie éthérique, vie nébulaire [Is this an ethereal life-form, a nebular one]?" They cannot conclude, but experience a phenomenon that reminds the reader of Kelvin's encounter with Solaris in Lem's novel of the same name: "Plusieurs colonnes s'étant heurtées au *Stellarium*, la phosphorescence s'arrêtait à partir de la paroi opposée; au reste, les segments communiquaient par des colonnes amincies qui contournaient notre abri. Comme normalement, les colonnes étaient droites ... il nous fallait admettre que la jonction s'était faite après notre arrivée [As several columns had collided with the *Stellarium*, their phosphorescence stopped when they reached the opposite wall; what's more, the segments communicated with each other by means of the thin columns that surrounded our shelter. As was normal, the columns were upright ... we had to admit that their junction had formed after our arrival]" (47). These entities simply pass around the human ship, then re-form on the other side. They do not harm it. They acknowledge its presence by adapting to its shape. But if they are sentient, they neither seek to understand nor communicate with the human object. Rosny's scientists conclude that, if these beings, like the other vegetative things they first observed, are forms of life, then "intelligent communication" is impossible with either.

Rosny's astronauts hope to find analogous, or in their words "homologous," forms of life on Mars. They are especially eager to find a "troisième plan de vie quelque part [a third form of life somewhere]." They soon learn that, in fact, Martians do everything in threes. When they land on the surface, in order to get a closer look at the "structures" they observed from the air, they discover an entire kingdom of life, some forms of which bear analogies to Earth plants and some to primitive animals. All the forms of what they call the "zoomorphes" prove, however, to be as Earthbound as the giant towers of light, which they name "ethereals," are beings that belong to the air. Moreover, all lifeforms in this kingdom appear to be organized according to a ternary schema. And, when the astronauts autopsy a zoomorphe, they discover a circulatory system that has adapted to Mars's lack of water, where life does not depend on the flow of liquids, but rather on what appears a "gaseous," or even solid, molecular movement. These beings, these "formes sans forme [forms without form]," seem to constitute a completely different kingdom of life from ours. What they do share with us, however, is their ability to adapt, chemically and physically, to a radically different environment.

The explorers soon discover as well that the zoomorphes' development comes, as was the case with the "ferromagnetics" in *La Mort de la Terre*, at the expense of another kingdom of life, the third life-form they were hoping to find. Throughout earlier encounters with various zoomorphes, the astronauts, despite striking differences, become increasingly aware Mars is not totally incompatible with their form of life: "Ce monde n'était plus (et combien mélancoliquement!) incompatible avec le nôtre. Les formes flexibles qui ondulaient sur le rivage et dans la plaine semblaient d'incontestables homologues de nos végétaux [This world was no longer (and how sad a discovery this was!) incompatible with our own. The flexible forms that undulated on the riverbanks and on the plains seemed unquestionably to be homologues to our flora]" (36). The third life-form, the Tripèdes, reveals itself, by the same process of homology, to be clearly humanoid. It "walks" upright, even though, as stated, its structural module is trinary instead of binary: "Dressées sur trois pattes, le torse vertical, elles (les creatures) avaient positivement quelque chose d'humain. Leurs visages même, malgré leur six yeux et l'absence de nez, leurs visages, dont la peau était nue, suggéraient je ne sais quoi d'homologue à notre espèce [Standing on three legs, their torsos upright, these creatures had something positively human about them. Even their faces, despite their six eyes and absence of nose, their faces whose skin was without covering, suggested something vaguely homologous with our species]" (61). The humans, in fact, are struck, not by these beings' monstrosity, but by their beauty, and (as an objective scientific assessment) by the apparent superiority of their organs to ours: "Aucun de ces grossiers appendices de chair que sont nos nez, nos oreilles, nos lèvres, mais six yeux merveilleuses...." [They have none of those vulgar appendages of flesh that we call noses, or ears, or lips, they have simply six marvelous eyes...] (61).

The Tripèdes capture Jean, but do not harm him. On the contrary, they prove to be simply curious about the visitors, and use Jean as means of learning about what they might be. Jean at once establishes communication with the Tripèdes, and relays information via Morse code to his companions, who are circling Mars's surface, seeking to know whether it is safe to land. Through Jean's account, they discover that there appear to be chemical differences between the species; for example, Jean is unable to eat their food or drink their water. The astronauts also learn that they live beneath ground. Instead of seeing them as savages, however, the scientists immediately understand that they are the victims of environmental change ("L'habitation sous terre implique l'appauvrissement de la planète [Their living underground is the result of the depletion of the planet]"). Their weapons prove to be quite sophisticated. But because these weapons lack power, they are judged to be "l'indice d'une civilization actuelle ou passée [an indication of a civilization on the wane or already past]" (65). Yet they are not seen as a decadent race but rather as a "devolving" ("décroissante") one, the victim of unavoidable evolutionary changes in the Martian environment. Reflecting on these changes, Rosny's astronauts see a problem that could very well be ours in some distant future — depletion of water resources: "Comme les hommes, ils appartiennent à une animalité dont la vie dépend d'un liquide.... Or, leur liquide, leur eau, est devenu rare ... et peut-être n'est-ce plus la même eau que jadis? [Like human beings, they belong to an animal species whose life depends on a liquid.... Their liquid, however, their equivalent of water, has become scarce ... perhaps it is not even the same sort of water they previously had]" (65). In Rosny's 1910 novel, *La Mort de la Terre*, he chronicles the death of carbon life on Earth due to the disappearance of water. Earthmen here, at a much earlier stage in their development, discover a Martian race dying of the same loss of their water. On Mars, however, this loss is not yet a catastrophe for us; it is more a warning, which could allow humanity to change its ways, and perhaps avoid this fate at least.

The Martian environment, then, seems to decree the eventual death of the Tripèdes and their kingdom of life. Diminishing water supplies favor the zoomorphes, whose circulatory system uses solid matter. In turn, zoomorphe activity on the ground is altering the chemical composition of the soil, making it unable to sustain Tripède food and life. As the astronauts establish communication with their Martian homologues, they find an evolved species in a struggle for survival with a well-adapted, ascending kingdom of life. The Tripèdes see themselves as losing this struggle. They are a highly intelligent species, which possesses the technology needed to repel the zoomorphe invaders. Yet they have apparently lost the will and ingenuity to use this technology effectively.

This Martian scenario broadens the vision that dominates all of Rosny's previous works. From the prehistorical evolutionary struggle of *Les Xipéhuz* (1887), through contemporary stories of mutations and alternate dimensions ("Un autre monde" [1895]), to the extinction of carbon life in *La Mort de la Terre* (1910), evolution concerned our species mainly, and Earth exclusively. The death of the Earth is the death of mankind's Earth, and yet there is another, iron-based form of life that awaits its evolutionary future on this same, ecologically transformed planet. In all of these works, as well, a sense of evolutionary fatality is suggested, despite the fact that Rosny's pluralist vision posits a continuity of evolving life-forms. Bakhoun, in *Les Xipéhuz*, defeats the Xipéhuz, an alternate kingdom of life engaged in a struggle of territorial imperative for Earth, in the manner of classic Darwinian survival of the fittest. Yet in the end he laments their destruction as a crime against life itself. Across the span of human time, however, in *La Mort de la Terre* it is now mankind's turn, as apogee of carbon life, to perish. Targ, the Last Man, both expresses sympathy for the successor life-form, the "ferromagnetics," and freely offers up "quelques parcelles" (a few elements) of his final carbon being to their primitive existence. We would like to think that, through this act of altruism, some significant aspect of human accomplishment might pass the evolutionary gulf, might become an important component of the development of an eventual ferromagnetic civilization. Even so, whatever the odds of this happening (and they are infinitesimally small), this remains small consolation for the death of mankind. The reader remains gripped by evolutionary fatality.

Rosny's voyage to Mars, though on the surface concerned with plural life-forms and parallel evolutions, offers a very different scenario. Space adventure permits what we might call a foreshortening of evolutionary time and process. For when they remained on Earth, humans could only follow their own evolutionary journey, over eons of time. But now, as a young, adventurous spacefaring species, humanity is free to intersect the evolutionary path of other "homologous" species. On Mars, it is the Tripèdes who have seemingly reached the end of the evolutionary line, and have become the fatalists. The humans at first describe themselves as observers. They witness the decline of the Tripèdes, but claim that they are not there to change the course of events. The fact is, however, that the conditions they encounter on Mars will force them to do just that. In order to survive on the planet, they must find ways to change the chemical composition of Martian food and water, in order that their metabolism can assimilate them. They also must convert Martian sources of energy. We remember that they brought with them only a three-month supply of energy, and will have to return to Earth if they cannot find more. Instead, they accept the physical challenge of Mars. They do so by the process later SF writers will call "terraforming," by transforming Martian elements so that they can sustain Earth forms of life. In doing so, they make Mars essentially an extension of their much younger world, which gives them license to use their energy and knowledge to help the Tripèdes set up defenses against the zoomorphes. Later, when the Tripèdes, under attack from the zoomorphes, prove unable to adapt to contingencies and are routed, the humans are able to step in and change the direction of the process. They do what John W. Campbell's time traveler will do in his seminal story "Twilight" (1936), written ten years after Rosny's novel: they instill in the expiring species a renewed sense of curiosity, giving them both will and a means of adapting to new situations. Unlike Campbell, Rosny does not see curiosity as a uniquely human trait, something that, when passed on to successor beings as in Campbell's story, in fact makes them human. Instead, for Rosny, it is a quality that waxes and wanes on any given evolutionary curve, be it human or Tripède. The curve has now waned for the Tripèdes, but the younger human species is in full flower, and can infuse life into an otherwise dying race. It is in an evolutionary context, then, that Rosny's astronauts give back to the Tripèdes the will to struggle, the desire to be resourceful, intervening to reverse their demise. The Martians recognize this: "De proche en proche, votre science mettra notre espèce à l'abri des invasions.... Les envoyés de la Terre auront sauvé leurs humbles frères de Mars! [Little by little, your science will shelter our species from future invasions.... The envoys from Earth will have saved their humble brothers on Mars]" (90).

As mentioned, Rosny's earlier evolutionary novels, from his tales of prehistory to his powerful "last man" narrative, *La Mort de la Terre*, offer a vision that not only predates the posthuman speculations of Bernal and Stapledon, but in terms of pluralistic rigor strives to avoid the specter of anthropocentrism that continues to haunt their work. For instance, in his story "Le Voyage" (1897), Rosny's explorers encounter a kingdom of intelligent elephants. In the later *L'Étonnant voyage d'Hareton Ironcastle* (1919), discovery is made of a society of intelligent trees. With the publication of *Navigateurs*, however, there is a radical shift in Rosny's vision. For example, terminal mankind, in *La Mort de la Terre*, cannot escape the iron law of evolution. Rosny's three astronauts, by the very fact that they can leave Earth, are given the possibility of reversing evolutionary conditions, if only in a localized area. In this novel, things are reversed; it is given now to the Martian Tripède to tell humanity that "tous les vivants ont leur fin du monde [all living things have their end of the world]." But with the voyage of our intervening astronauts, absolute evolutionary time yields to a relativist sense of evolution, where mankind is now able to "navigate" the field of evolutionary possibility. The astronauts now exploit a freedom of movement that allows humanity to put evolution in their favor.

This fact is confirmed by the Tripèdes themselves, for whom this otherwise impossible encounter of species offers not only an "awakening" of lost faculties, but a real possibility of changing their evolutionary path. The three astronauts, who speak of "affinités électives" between themselves, express a similar natural affinity for the Tripèdes. They are not afraid to use the word "anthropocentrisme" to describe this affinity: "'Anthropocentriste!' s'écria Antoine. 'Les Ethéraux, voire les Zoomorphes, devraient nous paraître bien plus passionnants! Ceux-ci ne sont qu'une manière d'équivalent des Terriens.' 'C'est vrai ... mais vous, au fond, qu'est-ce qui vous intéresse le plus?' ['Anthropocentrist!' Antoine exclaimed. 'The Ethereals, even the Zoomorphes, should be of much more passionate interest to us! These here are only partially equivalent to Earthmen.' 'That's true ... but you, deep down, what interests you the most?']" (65). In theory, the scientists may claim to find Martian life more fascinating than life on Earth. They may even posit that, because it is more varied, it appears to offer more evolutionary possibility: "Je le juge plus étonnant [your planet] ... Nous n'avons qu'une sorte de vie ... vous en avez trois! [I would say your planet is more astonishing.... We have only one form of life ... you have three]" (84). Their Martian interlocutor, however, answers that the contrary is true. To them, Earth is more interesting, for the reason that mankind has, at a young evolutionary age, gone into space, whereas they, over their long span of life, have never done so: "Combien la vie de notre planète sera plus courte que celle de la vôtre! Déjà l'âge rayonnant est passé ... et il ne fut jamais permis à nos ancêtres de franchir les abîmes de l'Étendue ... Trop petit et trop éloigné du Soleil, notre astre ne pouvait avoir une évolution comparable à celle du vôtre! [How much shorter the life of our planet will be compared to yours! Already the age of glory is over ... and our ancestors will never be allowed to navigate the void of Space.... Our planet, too small and too distant from the Sun, can never have an evolution comparable to your planet]" (84). Seen in the mirror of Mars, human beings are able to affirm the power of anthropocentrism, to assert that mankind's singularity may be more important in the balance of things than the plurality of life-forms. Before mankind's coming, the Tripèdes saw themselves declining; now they witness an awakening: "Et cependant, depuis votre arrivée, quelque chose palpite en moi, un étrange désir de renouveau ... l'aspiration vers une vie plus intense et plus vaste! [And yet, since your arrival, something stirs in me, a strange desire for rebirth ... an aspiration towards a more intense and vaster life!]" (84).

Rosny's novels invariably end with a paradigm shift, real or possible, that leads to a new evolutionary adventure. All carbon life perishes with Targ in *La Mort de la Terre*. Yet a tiny bit of his being passes to the new life-form. This passage is vectored by what is an act of sympathy for the successor life, however alien and hostile that life-form may seem. Targ's parcel is freely given, an act of Darwinian altruism, but there is no assurance it passes on as anything but unassimilable matter. *Navigateurs*, however, ends with a union of life-forms, a cross-species "marriage" between the human male and the Martian female, that physically produces, in the form of offspring, a regeneration of Tripède life.

The factor of evolutionary empathy may play a role here. But this time the empathy in question is an active trait of one of the three human astronauts. We remember that, of the three, the narrator Jacques is presented as a dreamer and poet, the Michel Ardan of the crew, a being who acts with his heart rather than his head. It is Jacques who feels a "strange sympathy" toward the Tripède female, whose name is Grâce. Her name is significant, especially in comparison with the name of Targ's woman of promise in *La Mort de la Terre*. The latter is named Êré. She is a blonde atavistic beauty, suggestive of heroic legend, an Eve to an Adam for whom there can be no regeneration, because the "era" of their race is over, destroyed by irreversible physical change. Grâce suggests predestiny over destiny. She and Jacques, in terms of human reproduction, are physically if not chemically quite incompatible. Yet, even though Grâce may seem grotesque by all norms of human beauty, she proves to be supremely attractive to Jacques, and this in a vitally physical manner, one that plays on his five senses with a sort of synesthetic charm: "Elle me frôlait; je sentis passer je ne sais quel fluide, plus ineffable qu'un parfum, plus évocateur qu'une mélodie. Je naissais à une vie singulière et charmante qui prolongeait l'image de Grâce dans le passé et dans l'avenir" [She brushed up against me; I felt some sort of fluid pass through me, more ineffable than a perfume, more suggestive than a melody. I was reborn to a form of life both unique and charming which diffused the image of Grâce throughout my past and future existence]" (83).

We are far beyond empathy in the evolutionary sense. What we have instead is a mutation where physical sympathy effects a telekinetic union, which in turn causes the literal fusion of two separate evolutionary lines. Jacques experiences a new, trans-species form of love: "Ces beaux frissons, ces ondes prodigieuses, comment les définir? ... L'idée que ce peut-être de l'amour, au sens humain, me semblait absurde et même répugnante.... Pourtant, c'est bien du désir que je ressentais auprès d'elle" [Those delightful shiverings, those mighty waves of feeling, how can I describe them? Any idea that this could be called "love" in the human sense seemed to me absurd, even repugnant.... And yet, it was indeed desire that I felt when near to her] (91). In the end, there is a union of the two, which proves to be sexual and reproductive. Jacques, however, describes his experience in a mystical language that seems to imply an act of "grace," as a force that intercedes and co-opts the process of evolution: "Une étreinte, rien qu'une étreinte.... Rien n'était plus. Tout disparaissait dans ce miracle qui semblait le miracle même de la Création" [An embrace, nothing but an embrace.... And nothing existed any longer. Everything vanished in this miracle that seemed to be the very miracle of Creation itself] (91). The reader must remember that Jacques is a poet, and he seems here, in his effusions, to be, more than ever, the poet rather than the empirical scientist. In the end, Rosny leaves us to ask an evolutionary question: Can such a "union of souls" in fact produce a physical offspring, which in this case must be the tangible product of a real mutagenic process?

Navigateurs gives no answer, and Rosny may have chosen to leave it this way. For though the promised trans-species offspring does appear in the sequel he wrote, *Les Astronautes*, Rosny did not publish this work in his lifetime. It saw print in 1960, attached to the Rayon Fantastique edition of *Navigateurs*. And read here, in the context of 1960s France, where a new form of SF strongly influenced by the American Golden Age was emerging, this sequel makes perfect sense. In *Les Astronautes*, Jacques brings Grâce and his Martian "family" back to Earth. The reader is now given an explanation for the birth. It is now known that Martians reproduce by what Jacques Sadoul calls "quasi-parthenogenesis," thus conception among otherwise physically incompatible species can actually occur. An offspring of two species and two evolutionary paths can in fact be physically born — but only if the birth takes place on Earth, and under the control of the human species ("Il suffit aux femelles de désirer fortement un enfant tout en pensant avec amour au mâle qui sera en quelque sorte son père"). We must not forget that we are now in the era of Heinlein's *Stranger in a Strange Land*.

And indeed, in Rosny's last science fiction work, a significant change of direction occurs, one which prefigures the supremely anthropocentric vision of a Heinlein and his fellow writers. The act of mental attraction that draws the Martian female is described as one of evolutionary co-optation: "Dans les limbes de l'inconscient, il me semblait qu'un monde fût en train de se construire ... les possibles jaillissaient de l'éternité créatrice, et je sentais le monde de Grâce rejoindre le monde obscur de mes ancêtres" [In the darkness of the unconscious mind, it seemed to me as if an entire world were taking shape ... possibilities burst forth from this matrix of eternal creation, and I felt the world of Grâce become one with the dark world of my ancestors] (78). This cross-species child, then, is born of a co-optive action similar to that by which Rosny's astronauts inspire in the Tripèdes the "étrange désir du renouveau" [strange desire for renewal] that leads them to eliminate the zoomorph threat. Likewise, it is due to their "anthropocentric" sympathy with the median Tripède species that the astronauts can invoke Voltaire's argument against superior species: The Ethereals are "plus subtils, sans doute! Moins exposés aux contingences brutales ... mais peut-être moins intelligents, après tout" [no doubt more subtle in their makeup! Less exposed to violent contingencies ... but perhaps less intelligent after all] (67). What Grâce experiences, in fact, is the experience of being possessed, and by means of this possession, of being drawn into the evolutionary stream of humanity itself. As Grâce describes it: "Je sens en moi renaître des souvenirs qui ne sont pas de moi-même, qui viennent du fond de nos âges, au temps où Mars aussi connaissait les eaux vivantes!" [I sense memories being reborn in my mind which are not mine, which come from the deepest depths of our past, from the time when Mars also had living waters on its surface] (81).

After several decades of SF in which space-traveling mankind has been extending its genotype to the far reaches of the universe, Rosny's novel and sequel, with its scenario of biological domination, may seem quite tame. We must realize, however, that he wrote his Martian narrative a decade before Weinbaum, and several decades before Asimov's and Heinlein's visions of mankind imposing its evolutionary destiny on other species. We must also remember that Rosny's space novel marks a radical shift away from the evolutionary fatalism of his earlier work; it appears, in fact, to mark a turn toward the aggressive anthropocentrism of the coming American Golden Age. Rosny's explorers (they openly compare themselves to Captain Cook, and even fear to suffer his fate among the Martian "natives") not only admit to an "anthropocentric weakness," but seem to glory in their short, dynamic existence. They lay claim to a powerful position at the normative center of life-forms, from which they are able to contain the mineral-like slow life of the zoomorphs, while at the same time ignoring the Ethereals, who are clearly ancestors of Heinlein's Martian Old Ones. In Rosny's final SF novel, mankind is already the small mote in God's eye: "Les hommes ne sont que des bestioles ... mais quelles bestioles!" [Men are nothing but tiny little insects ... but what insects they are!].

There remains one thing to consider: the strange title of this Mars novel, *Les Navigateurs de l'Infini*. When Rosny, in this novel, shifts emphasis from evolutionary fatality to anthropocentric centrality, he abandons his pluralist vision for a sense of mankind's place in the universe that is closer to that of the French rationalists and dualists. Clearly, his astronauts fly only to Mars, not into infinite space. But this small taste of the dark void between planets is enough to trigger their Pascalian meditation of the condition of mankind. Their real trial, perhaps, is not their adventure on Mars, but their encounter with the Pascal's vast infinity: "Ensuite des jours, plus lents, plus monotones, dans les abîmes noirs, dans le mystère éternel. L'Espace! Nous ne savons pas plus quelle réalité il dissimule que ne le savaient ceux qui crurent au vide ni ceux qui inventèrent des mondes à quatre, à cinq, six ... à n dimensions, pas plus que l'Eléate, que Descartes, que Leibniz ou notre Arénaut, conquérant de l'interstellaire" [And then we spent day after day, days ever slower, ever more monotonous, in the black abyss, in the eternal mystery of space! We have no idea what reality it conceals, no more idea than those had who believed in the vacuum, nor those who invented world of four, five, six ... or n dimensions, no more idea than Zeno of Elea, than Descartes, than Leibniz, or even our own Arénaut, conqueror of interstellar space] (42). On the evolutionary level, there is some degree of "selection," for it is Jacques who, unlike his companions, single with no family on Earth, is chosen to take the "next step" in human evolution. The larger frame of this novel, however, is metaphysical, ultimately dualist in the Cartesian or Pascalian sense. Here it is the mystery of *res extensa* that fuses the three astronauts into a common humanity as it faces the void: "La traversée de l'abîme interstellaire, l'isolement dans un astre perdu au fond de l'étendue faisaient de nous trois un seul être" [It was the act of crossing the interstellar void, the possibility of being isolated on some lost star at the end of the universe, that made the three of us into one single being] (62).

Even so, Rosny in this novel seems to be at a crux between Pascal and Heinlein, between on one hand the rationalist's sense of mankind's loneliness as sole rational being in the universe, and on the other, the idea developing in Golden Age SF that mankind derives its power precisely from this rational "edge" it has over empty, unthinking nature, despite its immense physical disadvantage. Both of these extremes on the Cartesian coordinates stand in sharp contradiction to Rosny's pluralist vision. We believe we see this vision operating on almost every page of this story, in the great detail with which he describes the different species on Mars, by which he delineates a complex

Martian ecological system. Yet the novel proves vastly different, when we compare it with an earlier work like "Un Autre monde" (1897). Here a human with mutated sense of sight perceives a "race" of beings in a parallel dimension, the Moedigen. He observes and describes them, again in great detail. But neither he nor the scientists he works with seek to enter their world to influence or co-opt it. In *Navigateurs*, however, Rosny's humans effect the fusion of species by which Mars will effectively be "terraformed." The evolutionary vision integrates mankind into a larger ecology of variables, and in doing so defines its role in relation to a larger play of forces. But the astronauts' experience of interstellar void in this novel sets mankind against nature, makes space and all in it things, *res extensa*, to be conquered, humanized. In this sense, Rosny's astronauts become the ancestors of the "tough critters," the striving monads of the later work of Heinlein and his contemporaries. In relation to infinite space, the astronauts take baby steps. But these steps offer, in turn, a taste of unlimited expansion. As space-age Captain Cooks, they work to escape the fate of their ancestor among the cannibals. In doing so, they receive the reprieve from destiny that Heinlein's protagonists will make into the building blocks of mankind's assault on the void, on (as Jerry Pournelle puts it) all the real estate that is out there for the taking. The distance between Rosny's evolutionary novels and his space adventure can be measured by making a comparison between Wells's *The Time Machine*, and the George Pal film remake of 1960, the same year Rosny's novel and sequel were published as a mass-market SF paperback. Wells shows us the inexorable and inevitable demise of mankind and his world. Pal shows us the Time Traveler, like Rosny's astronauts, teaching the Eloi to resist. The intervention of Pal's Traveler, like the intervention of Rosny's astronauts in the Tripèdes' destiny, resets the evolutionary clock in humanity's favor.

In conclusion, Rosny's space novel and its vision, coming as it does in a decade when the evolutionary fatalism of his earlier speculative works was just coming into vogue (Olaf Stapledon's *Last and First Men* was published four years later, in 1930), is all the more extraordinary for its vision, and for the fact that it prefigured a whole movement of American and British space age writers who could not have known it. By the time of its paperback publication in 1960, there was already a flourishing French tradition of postwar space operas, which were strongly influenced by the "globalizing" vision of writers like Heinlein, and in which Mars continued to play a crucial role, either as impediment to human expansion, or as jumping-off place for distant space explorations. There were numerous Mars novels of this kind in the Fleuve Noir Anticipation series, which began in 1953 and ended only in 2001. An example is Christopher Stork's [aka Stéphane Jourat, who wrote 48 novels for the series] *Made in Mars* (1985). Here a scientific space lab, with crew from all Earth's nations, is preparing a mission to Mars. Caught in a surprise attack at the beginning of a nuclear war on Earth, it is propelled into deep space, where it acquires the ability to travel in time. These new *Navigateurs de l'Infini* travel the void only to discover that Mars is their destination after all, that their sole enemy is a race of Martian humanoids. These are beings who do not need to invade Earth. They simply use their ability to time travel to control our evolutionary destiny, feuling wars on Earth across our bloody history, from prehistoric to atomic times. The novel moves through the Martian corridor, as does Rosny's work and numerous American space-time epics. The Martians are not only, as in Wells, Weinbaum and many others, our doubles. In Stork's case, they are actively reversing the work of Rosny's astronauts, "Martian-forming" our world in turn, in a kind of Pascalian *renversement du pour au contre*. In this novel, Rosny's brief fling at Heinleinian expansion meets once more with destiny. But this time it is not evolutionary determinism so much as Cartesian absolutism. The dualism between human mind and *res extensa* always holds. Neither history nor human nature can change. Indeed, Stork's novel ends with the Earth lab pursued by the Martian ship eager to destroy it because the humans know the secret of Martian warmongering. It lands in biblical Sodom, where they are destroyed as our history inexorably plays out, the Martian ship raining fire on the cities of the plain. Rosny's brief excursion into space exploration ends up the victim of the Cartesian mind games of French SF.

To this day, Rosny's novel has not been translated into English. Its rediscovery here, as an early vision of Mars that is parallel to, yet at the same time a harbinger of, the Mars of the American Golden Age, shows us that SF studies urgently needs a broader, international frame of reference. Without such a frame, there can be no accurate sense of the genesis of the genre. That sense lacking, there is no way to measure the cultural complexity that surrounds a theme such as space travel.

Works Cited

Asimov, Isaac. "Introduction: The Second Nova" *The Best of Stanley G. Weinbaum* (New York: Ballantine, 1974).
Rosny, J.H., Aîné. *Récits de science-fiction* (Verviers, Belgium: Marabout, 1975).
Sadoul, Jacques. *Histoire de la SF moderne: 1911–1984*. Paris: R. Laffont, 1984.

Two

The Uses of Mars

DIBS ON THE RED STAR
The Bolsheviks and Mars in the Russian Literature of the Early Twentieth Century
Ekaterina Yudina

Mars is a popular destination in Russian and Soviet literature. It is a place of import/export for the revolution in numerous works of fiction written in the first quarter of the twentieth century. It is an experimental natural laboratory only a space trip away from the U.S.S.R. in the popular culture of the 1950s and 1960s, with its promise that "the apple trees will blossom on Mars."[1] It is a destination for those escaping the Soviet reality in the 1970s. It is still one of the most bustling themes within modern Russian literature, and a recently published collection of stories by twenty authors, *New Martian Chronicles*, reflects this undying interest. Most of the texts stand in a dialogical relationship with the pioneering work of Russian SF, Aleksandr Bogdanov's novel *Red Star* (1908). In this chapter I compare it to Aleksei Tolstoy's *Aelita: The Decline of Mars* (1923).

One of the notable members of the Socialist-Democratic movement in Russia since 1896, Bogdanov was also a philosopher, economist, scientist, and founder of the world's first blood transfusion clinic. Working on his novel in the aftermath of the first Russian revolution, he populates his Mars with an exemplary communist society, describing in great detail its structure, labor principles, gender relationships, and language. Fifteen years and a whole new post-revolutionary era later, Tolstoy wrote his journey to Mars through the prism of an émigré liberal on the verge of repatriation. Although both novels use the symbolism of red as the color of the revolution, they treat the Martian theme very differently. Whereas *Red Star* belongs to the utopian genre and is driven by Bogdanov's revolutionary enthusiasm, *Aelita* is a parody of both SF and the love story (as well as Spengler's *The Decline of the West*). Hoping that his vision will intoxicate his readers, Bogdanov makes sure that all their potential questions are being answered. His journey to Mars is scientifically feasible and has a clear mission. Bogdanov gives lengthy descriptions of the ideal society, and his protagonist has the most sincere desire to learn from the Martian model. On the contrary, Tolstoy's main characters are goofy and do not have a clear mission when they start their journey to Mars. As for the social order they discover on Mars, it is far from being ideal (but neither is the society on Earth, even in socialist Russia).

The events of *Red Star* take place during the first Russian revolution of 1905. Leonid, the protagonist of the novel, is invited by a comrade from southern Russia, who turns out to be a prominent Martian named Menni, to join him on a trip to Mars to observe a utopian socialist society — perfect, rational, conflict-free. The Earthling communicates with the locals, falls in love with a young doctor, Netti, and studies the ideal social and economic world order established on Mars in the seventeenth century (the events of the Martian revolution are described in Bogdanov's second novel, *Engineer Menni*). The reader follows Leonid around the Red Planet, through the dwellings and workplaces of the Martians, learns about their egalitarianism, gender equality, sex life, labor structure, and so forth. Perfect as it seems to the Russian revolutionary, this society is not entirely problem free, since the harshness of the Martian environment requires, sooner than later, the colonization of another planet that would involve the destruction of its aboriginal population. Having learned that it is the Earth they are considering for colonization, Leonid kills Sterni, the main proponent of this idea. After that, Leonid is deported back to Earth, only to be willingly abducted again by his lover, Netti.

The name of the protagonist of Bogdanov's novel, Leonid, a Russian variation of Leonidas, from Greek λεων, "lion," is a significant choice. With its royal (both human and zoological) undertones, it also points to the parallel with the Spartan king, who, with only 300 soldiers, stopped the Persian invasion of Sparta, much as Leonid in *Red Star* prevents (or at least postpones) the Martian invasion of Earth.

The names of the protagonists of Tolstoy's *Aelita* also contain zoological references, but of a very different sort. That Tolstoy's novel is a parody is apparent in his choice of names: Los, in Russian лось, is "elk," whereas Gusev, from Russian гусь, is "goose." Sending his elk and goose on a Martian journey, Tolstoy cannot be further away from Bogdanov's careful selectionism of Earth's ambassador. The novel opens with the engineer and inventor from Petersburg, Mstislav Los, looking for a companion to go to Mars — to respond to some mysterious radio signals transmitted from there. Another, deeply hidden, personal reason for Los to leave Earth behind is the recent death of his beloved wife. His fellow traveler, an ex-soldier, Aleksei Gusev, also escapes his life, which has become too boring, with no combat prospects. No wonder that once on Mars, both of them pursue what is more dear to their hearts, rather than proudly claiming Mars for the Russians: Los falls in love with Aelita, the Princess, and Gusev joins the uprising. Still, this ill-fated uprising is an attempt to export the revolution — and in this, the two novels connect. In a reverse parallelism to Leonid, who goes to Mars on a kind of revolutionary internship, the main characters of *Aelita*, Los and Gusev, try to export the revolution from Petersburg into the galaxy. Though Los is reluctant at first, Gusev convinces him that it is not so much the actual revolution on Mars that they need as some bureaucratic proof of the revolution (another radical difference from Bogdanov's idealistic narrative). However, when the revolution actually starts on Mars, Gusev is only too happy to join in and lead it, no longer for the sake of bureaucracy but for the sheer joy of fighting on the side of the workers. When the uprising is crushed, the Earthlings manage to escape and to return home to Petersburg four years later. It is significant that Tolstoy deliberately uses the old, prerevolutionary name of the city, in spite of the fact that it is called Petrograd in 1921, when the novel starts, and Leningrad in 1925, when it ends.

Having a better understanding of the main characters chosen by the two novelists to represent the planet Earth, let us compare the two descriptions of the nature of the Red Planet and its social order. In Bogdanov's novel, Leonid is surprised to find the red vegetation on Mars:

> The substance which gives it this color is similar in chemical composition to the chlorophyll of plants on Earth and performs a parallel function in their life process, building tissues from the carbon dioxide in the air and the energy of the Sun. Netti thoughtfully suggested that I wear protective glasses to prevent irritation of the eyes, but I refused.
> "Red is the color of our socialist banner," I said, "so I shall simply have to get used to your socialist vegetation."[2]

Thus, the symbolism of the color red is being emphasized, and after that the protagonist loses almost all interest in the descriptions of the exotic locale (only mentioning that the sky is deep, dark green), and switches all his attention to what concerns him most, people and their relationships.

In Tolstoy's novel, the sky over Mars is deep, dark blue, whereas the main colors of the planet are orange (its terrain) and lilac (the shades). Some of the vegetation is compared to cacti and pine trees; the reader also learns about luscious gardens, purple, canary-yellow and silver. In Tolstoy's description, Mars lacks the color red altogether. The significance of this becomes apparent during the first night the Earthlings spend on Mars, when its sun sets and an incredible red star rises in the sky. This red star is the Earth:

> In the ashen sunset, low over Mars, a large red star appeared. It rose like an angry eye. For several seconds the darkness was filled only with its gloomy rays.... The gloomy star, rising, burned brighter. Reaching the shore, Los stopped and, pointing to the star, said, "Earth."[3]

In the dialogue between the two Russian novels, the meaning of the key phrase, "red star," gets paradoxically reversed, from Mars to the Earth. In *Aelita*, it is the Earth and not Mars that is the planet of victorious socialism, but Tolstoy clearly does not share Bogdanov's enthusiasm for the revolution, and his choice of language (гневный, angry, and мрачный, gloomy) testifies to that.

The Martians in both novels are humanoids who resemble the Earthlings, except for their huge eyes. As Bogdanov puts it, "[the] appearance was highly original, deformed even, but not to the point that it could be called grotesque."[4] This deliberate desire of both writers to keep their fantasy on a short leash when it comes to inventing the new race could be explained by "the rules of engagement" in both romance literature and propaganda pamphlets, where emphasizing the differences is just too destructive.

Many pages of *Red Star* are devoted to the labor structure and productive process. A passionate anti–Taylorist, Bogdanov creates fully automated factories where workers are no longer appendages to machinery. Workdays are short, jobs are rotated, and all work is voluntary. As Richard Stites writes,

> Planning, productivity, labor discipline, and recruitment — all problems of developed industry outlined in the novel — became issues of heated debate among Soviet planners of the 1920s and 1930s. No wonder that Bogdanov's novel was sometimes invoked at the dawn of the First Five-Year Plan by economic chieftains and planners.[5]

Unlike Bogdanov's classless society, the members of the working class on Tolstoy's Mars live and work in old, dusty buildings, submerged into a hopeless, antlike existence, in contrast to the somewhat thriving bourgeoisie. But the society, the whole Martian civilization, is declining, and whether or not it can be rescued by the Earthlings is the topic of fierce discussion. An important component of this discussion is the possibility (metaphorical? sexual?) of the transfusion of hot, human blood to the more cold-blooded, intellectual Martians. The ruler of Mars, Tuskub, is categorically opposed to the idea:

> Hope and depend on the aliens from Earth? Too late. Pour fresh blood into our veins? Too late. Too late and cruel. We will merely prolong the agony of our planet. We will only increase the suffering, because inevitably we will become the slaves of the conquerors.[6]

The theme of mutual blood transfusion is one of the most important features of the Martian lifestyle in *Red Star*. Netti explains it to Leonid: "Quite in keeping with the nature of our entire system, our regular comradely exchanges of life extend beyond the ideological dimension into the physiological one."[7] The blood of the younger and/or healthier person regenerates and to a certain extent rejuvenates another organism, but a young person does not age in turn because "the age and weakness in the blood are quickly overcome by the organism, which at the same time absorbs from it many elements which it lacks."[8] In 1926 Bogdanov founded the Institute for Blood Transfusion where he applied these Martian principles and enjoyed the jealous patronage of the aging Bolshevik leaders who readily invested in every aspect of gerontology. Two years later, after numerous successful transfusions, including eight carried out on himself, Bogdanov exchanged blood with a young man suffering from both malaria and tuberculosis. Ever since, we the Earthlings consider Bogdanov dead, but perhaps that was just another way for a Martian to return home?

As for Aleksei Tolstoy, he too returned to his homeland. In 1923, following the example of his characters, Tolstoy repatriated to the U.S.S.R., accepted the regime, became one of the most popular Soviet writers, witnessed the success of a film version of *Aelita* and enjoyed another twenty-two years of privileged life as the "red count."

Notes

1. The song "The Apple Trees Will Blossom on Mars" was written by Vano Muradeli and Evgenii Dolmatovsky in 1963.
2. Alexander Bogdanov, *Red Star: The First Bolshevik Utopia*. Bloomington: Indiana University Press, 1984, p. 60.
3. Alexei N. Tolstoy, *Aelita*. New York: Macmillan, 1981, p. 43.
4. Alexander Bogdanov, *Red Star*, p. 31.
5. Richard Stites, "Fantasy and Revolution: Alexander Bogdanov and the Origins of Bolshevik Science Fiction," in Alexandr Bogdanov, *Red Star*, p. 7.
6. Alexei N. Tolstoy, *Aelita*, p. 109.
7. Alexander Bogdanov, *Red Star*, p. 86.
8. Ibid.

Works Cited

Bogdanov, Alexander. *Red Star. The First Bolshevik Utopia*. Bloomington: Indiana University Press, 1984.

Stites, Richard. "Fantasy and Revolution. Alexander Bogdanov and the Origins of Bolshevik Science Fiction" in Alexandr Bogdanov, *Red Star*.

Tolstoy, Alexei N. *Aelita*. New York: Macmillan, 1981.

THE MARTIANS AMONG US
Wells and the Strugatskys
George Slusser

"Mars with its reddish glow, seems a place
Of neverending wars of defeat"
— John Hatch

As with England, France and the U.S., emergent Russia at the end of the 19th century felt the lure of Mars. As planets go, Mars was a visible neighbor. Moreover, with its "canals," we thought, it could harbor a technologically advanced civilization equal, or superior, to our own. It was a place we might invade; or, as H.G. Wells graphically demonstrated, it might invade us. The return to order at the end of *The War of the Worlds* (1898) seemed to calm the fears of Martian invasion. Even as World War I raged, Mars became the terrain of John Carter and fantasies of adventure and empire. On the eve of World War II, however, Orson Welles reawakened the fears of an invasion from Mars. The lesson this invasion teaches is that of a great power unprepared, caught by surprise, totally run over.

This is a lesson that resonated particularly with the postwar generation in the Soviet Union. Even in darkest Stalinist times, Wells's works remained available to Soviet readers, most likely because of his vision of one-world socialism. The coming of age of this postwar generation coincided with de–Stalinization and the Khrushchev "thaw." And in the work of the two major Soviet science fiction writers of this period — Arkady and Boris Strugatsky — we see open and sustained fascination with Wells's Martians. Alien invasion would seem a theme of primordial interest to a culture so often invaded. The Strugatsky canon is particularly rife with flashbacks to scenes of literal Nazi invasion. Scenes of futile Russian cavalry charges against the iron juggernaut of German tanks remind the reader of human assaults against impregnable Martian weapons in Wells's novel. But the impact of Wells's Martians goes much deeper in the Strugatskys' work, forming a subtle network of intertextual references that point to an ongoing cultural dialogue between Wells's work and Soviet society.

For Wells, an essential aspect of alien invasion is social disruption. At the deeper level of his novel, Wells asks the question: Who are these Martians who invade, who randomly destroy while obeying a logic that totally ignores human suffering? "Cool intelligences" that look upon humans as we do ants, that blindly and "rationally" execute their plan without the least regard for the beings they control, these Martians, for Wells's narrator, could be a future form of humanity itself, where reason has replaced heart. For Soviet writers like the Strugatskys in the 1950s and 1960s, the Martians could represent the Stalinist planners of the Soviet state. At the very least, in a society that preaches dialectics, Wells's Martians offer a dialectical challenge to official Marxist doctrine, which certainly survived Stalin's fall. Soviet Marxism promises the perfectibility of humanity through the march of dialectical materialism toward a utopian future. Wells's Martian invasion reveals the very opposite. For the subject of that invasion is less the Martians themselves than the impact of Martian actions on human beings in terms of their fears, cowardice, venality, futile heroism. In a culture where the march to the future is seen as implacable, Wells's Martians reopen the question of human imperfectibility. For in Wells's novel, human beings are both their own oppressors and their own oppressed. This is a message the architects of Soviet ideology did not want to hear. But hidden behind the mask of Martians, it can be sneaked in. This essay examines the role Wells's Martians play in shaping the Strugatskys' science fiction.

Mars in Russia: The Theme

Mars the planet has the perfect color for an ideal Soviet location. The prerevolutionary physician and social philosopher Alexander Bogdanov had already set his workers' utopia on Mars in *Red Star* [Krasnaia zvezda] (1908). Later, Mars becomes the center of revolutionary activity in Alexei Tolstoy's *Aelita, or the Decline of Mars* (1923), followed by Yakov Protazanov's film *Aelita* (1924). We find a very different Mars in this space drama. After the glorious revolution, engineer Los travels with former soldier Gusev to Mars in a rocket. They find human beings, survivors from the fall of Atlantis. These have blended with "native" populations, and have set up an advanced capitalist society where, in the manner of Fritz Lang's film *Metropolis* (1926), the working class is oppressed by their bosses. In addition Mars, in the manner of Wells, is a physically declining world. Gusev mounts a workers' revolution that fails; he and Los, who has fallen in love with Aelita, leave Mars and return to Earth.

The novel and film are usually read as presenting aspects of doctrine in a time of forces contending for power. But in a sense, both Tolstoy's novel and the film are primary examples of early Russian science fiction, a genre seen by Lenin, for example, as a powerful means of conveying ideas. Until Stalin and the doctrinaire bolsheviks banned SF in the 1930s, the form thrived. Science fiction was, during this period, an international form. For example, the influence of Edgar Rice Burroughs on *Aelita* is clear. Mars is already a dying world in Burroughs's *A Princess of Mars* (1912). In Russia, during the turbulent decade of their writing, Burroughs's Mars novels were extremely popular and rapidly translated.

These novels belong to a subgenre of the developing form of science fiction — the space opera, a form emerging around Wells's *The War of the Worlds*, in France, England and the United States. The earliest examples of space opera may in fact be French, in the form of the visionary "voyages" that precede later tales of true space travel. For example, Restif de la Bretonne, in *Les Posthumes* (1802) is already detailing Martian customs. C.I. Defontenay, in *Star, ou psi de Cassiopée* (1854) offers his readers actual space opera, adventures in space. In Henri de Parville's *Un Habitant de la planète Mars* [*An Inhabitant of the Planet Mars*] (1865, published by Hetzel the same year he published Verne's *De la terre à la lune*), activity in outer space brings a Martian to us, in the form of the intact remains of a "man from Mars" brought to Earth by a comet, proving that there is life on Mars, and that this life resembles human beings. These works were followed by the Martian cycle of Camille Flammarion. Flammarion's *L'Astronomie populaire* [*Popular Astronomy*] (1879) was widely read and translated, certainly in Russia, where the second language of the educated classes was French. Flammarion mixes mysticism (reincarnation) with detailed descriptions of Martian life-forms in his aptly named *Les Terres du ciel* [*The Earths of the Sky*] (1884). He continues his speculations on what we would call today the Martian ecology in *La Planète Mars* [*The Planet Mars*] (1888), in which Martians, as reincarnated humans, are evolving toward the perfection of the human race. He further speculates on Mars as chosen place for human reincarnation in *Uranie* (1889). Finally, in *La Fin du monde* [*The End of the World*] (1893), Flammarion gives Mars the role of sending the message that warns us of an impending comet strike on Earth.

On the more popular, "pulp" side, we have such immensely popular works as Henry de Graffigny's and Georges Le Faure's *Les Aventures extraordinaires d'un Savant russe* [*The Extraordinary Adventures of a Russian Scientist*] (1889). In Volume 2, "Le Soleil et les petites planètes ["The Sun and the Smaller Planets"]," the team of French and Russian scientists who are exploring the solar system descend from Phobos to the surface of Mars, where they (quite predictably) encounter a technologically advanced civilization of humanoid creatures with wings. Wells, in his story "The Crystal Egg," written in 1897, one year before *The War of the Worlds*, describes his Martians as winged creatures: "The air seemed full of squadrons of great birds" (672). Graffigny's and Le Faure's volumes were handsomely illustrated by Albert Robida and other artists, who gave striking visual presence to landscapes and alien creatures such as the winged Martians. In the same pulp vein, we have works like Arnould Galopin's *Le Docteur Oméga: Aventures fantastiques de trois français dans la planète Mars* [*Doctor Omega: The Fantastic Adventures of Three Frenchmen on the Planet Mars*] (1905), and Gustave Le Rouge's *Le Prisonnier de la planète Mars* [*The Prisoner of the Planet Mars*] (1908) — Burroughs *avant la lettre*. During World War I, we have works like André Mas's *Les Allemands sur Vénus* [*The Germans on Venus*] (1914), in which (in a manner that foreshadows Heinlein's Nazis on the Moon in the post–World War II *Rocket Ship Galileo*) the Germans take Venus, Russia gets the Moon, and France Mars. France's pulp Mars continues well beyond World War I, indeed beyond World War II as well, with numerous Mars novels appearing in the 1950s in Fleuve Noir Anticipation. Notable in the early 1920s is Théo Varlet's *Les Titans du ciel* [The Titans of the Sky] (1921), which openly provides a sequel to Wells's *War of the Worlds*. The novel continues Flammarion's interest in reincarnation, this time with Martian "souls" invading bodies on Earth.

The Anglo-American pulp novel also flourished during the decades before and after the publication of Wells's masterpiece. In the wake of Schiaparelli's observations and new map of Mars, a number of popular novels saw Mars as the perfect place for a utopian society. Examples are Percy Greg's *Across the Zodiac* (1880), Hudor Genone's *Bellona's Bridegroom: A Romance* (1887), and, with a nice Edwardian twist on the problem, Ellsworth Douglas's *Pharaoh's Broker* (1899), in which a financial speculator travels to Mars and finds there a society identical to that of Pharaoh's Egypt. The pharaohs have conquest in common with Wells's Martians. Garrett P. Serviss, however, shows American ingenuity taking the fight to the tyrants in *Edison's Conquest of Mars* (1898), a hastily produced "sequel" to Wells's novel. Following changes in publishing media in the early decades of the 20th century, the romances of Mars and interplanetary war novels became space operas, and were increasingly serialized in often lurid pulp magazines. Burroughs' Mars novels, for instance, were serialized in *All-Story Magazine* before publication in book form. It is these serialized tales that Evgeny Zamyatin most likely read during his stay in England. When he returned to post-revolutionary Russia to write his masterpiece *We* (1921), Zamyatin revealed that he read these pulps. In fact, in *We*, it is precisely the pulp spirit of unruly rebellion that becomes the enemy of the One State. This spirit has not only penetrated the Taylorized mind of D-530, the narrator, but of the official organ of that state, the *One State Gazette*. Here is how the gazette describes the Integral: "You will integrate the infinite equation of the universe with the aid of the fire-breathing, electric, glass *Integral*. You will subjugate the unknown beings on other planets" (Zamyatin 1). On page 1 of this narrative, the totalitarian state (read Bolshevik Russia) reveals it is already fatally infected with the rebellious spirit of science fiction and its vivid imagery of Martians and beings from other worlds. Wells's *War of the Worlds* is the seminal text here, and the model for future alien invasions. Indeed, it is these same Martians, now enhanced by a long series of sequels in print and film, that will be brought to bear, increasingly, by the Strugatskys on the postwar Soviet Union.

Wells in Russia: The Vector

It is clear that Wells has remained a widely read writer in Russia and the Soviet Union through thick and thin. This is especially true for his science fiction, of which the best known of all is *The War of the Worlds*: "H.G. Wells and his novels, especially his science fiction, have been part of the experience of the Russian reading public since his first works were translated and published in Russia at the end of the 1890s. One may be sure that every Russian reader more or less attracted by science fiction knows Wells's work."[1] Wells visited Russia and the Soviet Union three times, once before the revolution and twice afterwards. His attitude toward the development of society in the Soviet Union between the world wars remained ambiguous, to say the least. The important thing, however, is that his works, especially the scientific romances, remained in print, while works of Huxley, Orwell, Zamyatin and other dystopians were banned. Wells fascinated Zamyatin, who translated *The Time Machine* and other works into English. Bulgakov was loosely inspired by *The Island of Doctor Moreau* in his *The Heart of a Dog* [Sobach'e serdtse] (1925). Yuri Olesha, in *Envy* [Zavisti] (1927) interprets *The Invisible Man* as a parable describing the woes of the artist in a rigidly materialist culture. The James Whale film of *The Invisible Man* (1933) was widely shown in the USSR. During the bleak 1930s and 1940s Wells remained a threshold writer in the Soviet Union, essentially by virtue of his science fiction.

Interestingly, the George Pal film version of *The War of the Worlds* (1953) appeared the same year as Stalin's death. The twentieth party congress, in 1956, began the process of de–Stalinization. *Sputnik* was launched the following year, and Stalin's body was removed from the Kremlin in 1961, one year after the appearance of another Pal film, *The Time Machine,* in 1960. Julius Kagarlitski published *The Life and Thought of H.G. Wells* in 1963 (translated into English in 1965 by Wells's mistress Moura Budburg). Wells's presence was felt during this turbulent time. But this was also the time of the emergence of a new kind of Soviet SF, which in turn (no doubt because of the space race) involved a dialogue, or dialectic, with postwar American SF. The landmark work here is Ivan Yefremov's *Andromeda Nebula* [Tummanost' Andromey] (1958), translated as *Andromeda: A Space Age Tale*. The future utopian society may be orthodox Marxism; however, there are many elements of postwar American space exploration SF: generation starships, battles with cosmic aliens — one could call it a Soviet version of the first generation *Star Trek*. Yefremov's next SF work, *Heart of the Serpent* [Serdtse zmei] (1959) in fact is presented as a Marxist response, specifically, to Murray Leinster's benchmark story "First Contact" (1945).

One could say then that, at this time, there are other visions of Mars than Wells's moving through the net of Soviet censorship. These are either accepted as innocent-seeming "juvenile" fantasy or rejected as ideological straw dogs. Pal's film *The War of the Worlds* is an example of the latter, for it turns Wells's classic effectively into an expansionist tract, where (American) humans defeat the Martians and promise to take their planet sometime in the future. On the other hand, works like Heinlein's *Red Planet* (1949), and especially Ray Bradbury's *Martian Chronicles* (1950), appeared harmless enough, and could easily be turned into "models" for a new space race vision of Mars. In these works (as in *Andromeda*) expansionist Earth science turns the tables on Wells's feared Martians; Mars is an old planet, but now a planet with much less fearful aliens. In Heinlein, the Martians are an ancient culture; they still have tremendous physical power, but lack the youthful desire to use it; they ponder while Earthmen act. Bradbury's Martians are both decadent and dying, and they are invaded by what are clearly Americans. In a neat twist on Flammarion and French space opera, the essential Martian weapon used here against the imperial Americans is telepathy and reincarnation. It is in this context, at the debut of their writing careers, in the late 1950s/early 1960s, that the Strugatsky brothers first encounter Mars, via Bradbury and Heinlein. Even so, Wells's Mars is still very much alive, and will prove to be increasingly relevant to their growing vision of things. The Strugatskys will soon abandon the space-age Mars; they will come to take a deeper look at Wells's Martians. For Wells's Martians are, as the author clearly pointed out in the late work *Star Begotten* (1936), not only among us but within us as well. Bradbury's characters, in the final story of *The Martian Chronicles*, look down in the waters of Mars and see their reflections: They are Martians by default. Wells, from the first lines of *War of the Worlds*, lets us know that we have Martians embedded in our very beings.

The Mars of the Thaw: Wells Between Yefremov and Bradbury

In 1962, the Strugatskys published a curious "juvenile" SF novel, *Stazheri* (from the French *stagiaire*, literally an apprentice for a trade), translated in English as *Space Apprentice* (1981). It is ostensibly the story of a young welder, Yura Borodin, who misses his ship to the outer planets and, through a serendipitous meeting with famed engineer Ivan Zhilin, finds himself taken as "apprentice" on the command ship of Vladimir Yurkovsky, inspector general of the International Administration of Cosmic Communications. This seems, at first glance, a replay of Heinlein's *Starman Jones* (1954). From the book's prologue, however, we see that the focus is not, as with Jones, on Yura's rise from obscure youth to key player in the exploration of space. It is instead the story of the final voyage of Yurkovsky, Bykov and other figures who rose to prominence in the earlier days of primary exploration, and who now both are reduced to administrative duties. The pioneering spirit of the early explorers is beginning to unravel, both due to basic human failings and, more specifically, to capitalist greed. Yurkovsky puts a temporary lid on several problems. In the end, however, off the rings of Saturn, in sight of what seems a first contact with aliens, Yurkovsky and another old comrade perish. Yura is left at his destination. The homecoming instead is for Bykov, a broken man, but one who is still yearning for space. The real lesson is learned by Zhilin: "'The most important thing always stays on Earth, and I will stay on Earth too, I've decided,' he thought.... The important thing is to be on Earth" (231).

There is a strangely Wellsian ring to this novel. Yura's voice is one of over-enthusiasm, followed by ineffectual collapse. Yurkovsky's scope is the opposite; his vision is unsustainable when confronted with physical limits. Zhilin is the *via media*, yet like the narrator of *War of the Worlds* and other scientific romances, he returns to normal existence on Earth an alien, a stranger in a strange land. Interestingly, too, given the Wellsian tone of this novel, a sizeable part of the action takes place on Mars. Even so, Mars here, at first glance, seems to be the world of Bradbury even more than the Dying Planet Wells's narrator sees from afar. Humans are colonizing Mars. Involved, however, in day-to-day activities, and in perpetual struggle with giant "leeches," they have not taken the time to ask key questions as to where they are and who was there before them.

One figure only, the Pathfinder Felix Rybkin, has studied the land and its mysteries, and is able to point out, by measuring time in terms of a slow-blooming Martian flower, that what is called Old Base was not built by earlier settlers, but is in fact a Martian structure. What seems cement is not cement. The leeches that hide in caverns beneath this place attack two-legged humans because the original Martians must have been bipeds. The focus of the action now dissipates: Yurkovsky arrives, a great "roundup" of leeches is carried out, Yurkovsky shows (foolish) heroism by going into the caverns, Yura breaks down. In the end, however, the same complacency reigns as before — the humans return to their preoccupations. The Wellsian menace remains, however. The Martians are still out there, and Rybkin alone is willing to go out at night to scout for them.

Mars plays a special role as well in the story cycle *Polden' XXII vek* (1962, revised 1968) translated as *Noon: 22nd Century* (1978). The first story of the Bradbury-like chronicle of interlocking tales and characters is simply titled "Night on Mars." It recounts the trek of two doctors and two Pathfinders through the Martian night to deliver the first baby born on the planet — the first "man from Mars." Along the way they are beset by a monstrous creature that apparently attacks only standing humans. The assumption, again, is that the "original" Martians were bipeds. In the story, however, we never see a Martian. And Mars here is but a prelude to space. For in the stories that follow, mankind expands its reach, both in terms of scientific discovery and intergalactic exploration. Yet just as, in this first story, the real Martians are never seen, so mankind continues to experience a string of such "near-misses" along the way to the 22nd century. With only 98 percent of academician Okada's biomass downloaded, the immortality project fails. An elaborate experiment in telepathy, on the verge of breakthrough, suddenly goes silent. "The planet with all the conveniences" turns out to be a "biological civilization," an encounter for which no human first contact directives prove even remotely adequate. In the penultimate story, "The Meeting," two old comrades reminisce about the old days of Mars. They invoke the promise of biped Martians; they now ponder something greater — a tragically botched encounter, this time with what seems to be a genuinely humanoid alien. Hunter Pol Gnedykh shot this being on Crookes' Planet; its remains have been transferred, as a specimen, to a museum of exobiological species on Earth. In this story, the Strugatskys seem to take Yefremov's official theme — successful first contact with aliens who, if they are sentient, have to look like us — and turn it on its head with a touch of Wellsian irony.[2] Pol shot the creature because it does not conform to the human norm. But his old friend the taxidermist Kostylin knows better, and in this museum to the glory of human expansion, he traces the word *sapiens* in the dust under the specimen's inscription, and hastily erases it. Instead, the Strugatskys seem to echo Wells's Martian problem: which says that if the alien is always somehow *us*, it is because that alien remains a projection of our unregenerate, violent nature — we are and always will be the Martians. In the shadow of Yefremov's official doctrine of perfectibility in history, Wells's sense of an unregenerate human nature resurfaces.

The final story, "What You Will Be Like," contains Gorbovsky's embedded tale of the miraculous visit of a "future" human, Petr Petrovich, to his derelict space vessel. Petr Petrovich heals a fatal wound with a massage, climbs into the atomic reactors, restarts them, and plots the ship's homeward coordinates. At the end of *War of the Worlds*, Wells's narrator ponders the possibility that we may become Martians in the future. This implies that our technologically advanced future may only be another stage in our violent past. In a manner that suggests Wells, Petr's description of human progress — what we will be like — is oddly circular: "Just remember, if you are what you plan on being, then we'll become what we are. And what you, accordingly, will be" (Noon 318). What is more, if this visitor from our future claims the right to intervene in our progress, he admits that these interventions can (like Wells's Martian invasion) turn destructive: "We could goof something up and turn history head over heels" (Noon 318). The sense here is that what we will be like is what we always were and are — violent natures who resist the historical march to utopia.

What justifies Petr's claim to the right of intervention are vivid descriptions of juggernauts and war machines like those of Wells's Martians: "It's all right to intervene the way I'm doing now. Or like another friend of mine. He ended up in one of the battles near Kursk and took it upon himself to repel a German tank attack. He got himself killed, chopped into kindling" (Noon, 318). Slavin, the first child born on Mars, now speculates on Lenin's idea of the development of the human race: "You see, the human race began with communism and it returned to communism, and with this return a new turn of the spiral begins, a completely fantastic one" (Noon 319). Minus the communism, this could be Wells's narrator, after the Martian invasion, predicting a new, fantastic turn of history about to begin. Wells's narrator focuses on near events, on humanity's incorrigible nature and the future holocausts it will surely bring. Despite its final flourish of obligatory optimism, Slavin's message has a similar ring: "We gather strength for the future.... We will hear the approaching footsteps of the element of fire, but we will already be prepared to loose waves upon the flame" (Noon 318).

There is a segué here, from Petr Petrovich's right of intervention in history to a more overt use of *War of the Worlds* in the next major work, *Trudno byt' bogom* [Hard to Be a God] (1964). We remember the survivor-narrator at the end of Wells's novel, pondering the possible benefits of the just-"defeated" Martian invasion. In light of the human failings just witnessed, however, his utopian vision of a "commonweal of mankind" that will take the Martian "gifts" to human science and use them to carry that vision to other worlds, seems but another poignantly ironic example of blindness to the problem of human nature. The Strugatskys, with a bitter irony new to their work, seem to be putting the narrator's program to the test in this novel. The Martians intervened in our world. Now humans claim the right, with no doubt equally good intentions, to do the same in other worlds, where they are equally not wanted. As their inverted Martian invasion plays out, the Strugatskys clearly have not forgotten the deep sense of unregenerate human nature that, in Wells, underlies all such wishful visions of high moral "progress."

Anton, a scientist from the Institute of Experimental History, is part of a team sent to the planet Arkanar, where an alternate human society has reached the stage of our Middle Ages. The invaders this time are not bellicose Martians. They are the same "progressive" humans envisioned by Wells's narrator, armed with utopian ideology and superior science, who come with the "good" intention of allowing Arkanian society to leapfrog the bloody period of capitalist class struggle and pass directly to a state of socialist equality. Ironically, however, just as Martian intentions were never fully known to mankind, so the future humans remain misunderstood and unwelcome on Arkanar. Anton's efforts meet with unbending resistance to change in his opponents, a resistance manifest in cruelty, violence, and finally devastating war. Finally, pushed to the breaking point, Anton's own Martian self explodes beneath the veneer of super-civilized human, and he responds with deadly violence.

There is a clear parallel here. Wells's invaders are violent. Their acts, however, summon in turn the violence innate in human nature that lies beneath the "civilized" manners of the average Englishman — a violence the Martians only raise to a higher power. The optimism of Wells's narrator is short-lived. Likewise, following Anton's optimistic arrival and subsequent alienation from this world he has come to change, he returns to revisit the idyllic moment of the prologue, a stroll he and the other young future invaders of Arkanar took in an Earth forest before leaving for space. At that time, he ignored the implications of his discovery of the skeleton of a German machine gunner, chained to his weapon at the end of a forest road. After Arkanar, Anton returns to the forest to find this same road, which he travels again, but this time alone. Even though he now calls this road "anisotropic," the one-way road of history, he has learned that its end (Arkanar) like its beginning (the gunner) are marked by the same senseless violence. To travel this road is to return, like Anton, to the point of departure, but this time as a shade, as one powerless to change the terrors that lurk behind all promises of technological progress. Wells's narrator has a similar final vision: "I sit in my study writing by lamplight ... and feel the house behind and about me empty and desolate. I go out to the Byfleet Road ... and I hurry again with the Artilleryman through the hot, brooding silence. Of a night I see the black powder darkening the silent streets, and the contorted bodies shrouded in that layer" (*War of the Worlds* 189).

The Second Martian Invasion

In these early Strugatsky works, we detect Wells's Martians stirring behind the facade of official Soviet doctrine, as set forth in the SF works of Yefremov. If, according to that doctrine, the only form of life capable of higher evolution is *Homo sapiens*, and the only form of evolution is human progress through the spirals of history toward the classless society, then the role of Martians in the Strugatskys' work, insofar as they are images of our future selves, is to reveal the degree to which human nature remains refractory to such official progress and change. For Wells, these denizens of an old planet represent the capacity of our future to wage war on our present. It is a supremely unhistorical vision, one that is glimpsed over and over in the scenes of apocalyptic warfare that haunt chronicles of the future like *Noon: 22nd Century*, one that makes it hard for humans to be gods.

There is, however, the other side of Wells's Martian invasion. This is how we react to them, a process that leads to eventual revelation that we are Martians as well. *The War of the Worlds*, despite its alien Martians, is a narrative focused on human reactions. The narrator presents an anatomy of failings and, to a limited degree, of qualities, as human beings are subjected to external pressures which test the limits of their capacity to resist, and finally endure, when survival is the sole hope for any human future at all. Wells details the responses of a cross-section of "average" humans to Martian invasion. In doing so, he exposes the institutional and cultural complacency that may in fact be responsible for summoning the invasion. This is the lesson the Strugatskys seem to draw as they now, for the first time, openly name their source in the satirical "rewrite" *Vtoroe nashestvie marsian* [The Second Invasion from Mars] (1968, translation 1974). Wells's portrait of a smug and secure England at the turn of the century is transposed to a conventional Russian setting. The setting in fact appears to be pre–Soviet. It is in fact a traditional Russian village where the names of its denizens have been absurdly camouflaged with names from Greek mythology. In an ironic bow to Wells's English village, the Strugatskys' Russian village is in the process of being invaded, this time by "capitalist" Martians.[3]

This "sequel" to Wells's novel is also told by a narrator/witness/actor; here, however, he is a consummately obtuse and unobservant figure named Phoebus Apollo, who is surrounded by other aptly named figures such as the septic tank cleaner, Minotaur, and a policeman named Polyphemus (a possible reference to Wells's "The Country of the Blind"). Wells's narrator knows enough to chide his contemporaries for not seeing signs of the impending invasion: "Men like Schiaparelli watched the Red Planet ... but failed to interpret the fluctuating appearances of the markings they mapped so well" (War 13). Apollo, however, like the majority of his peers, never realizes there has been an invasion. These Martians are never noticed, for they simply encourage the townspeople to go about business as usual — the petty routine of petty lives.

The Strugatskys' Martians have learned from past mistakes. What they have learned is human nature and how it works. They have learned the fact that humans, when nudged in the right direction, will defeat and enslave themselves without external coercion. Their method is covert, a thoroughly modern form of co-optation through misinformation and propaganda. Apollo, a pensioner and philatelist, goes about his business, mechanically recording facts, in which the reader gradually sees the nature and extent of the Martian takeover. Articles appear in the local paper contradicting scientific evidence that there is no life on Mars. Patriotic calls are issued, inciting all citizens to keep their gastric juices healthy for the greater good. Local wheat crops are mysteriously replaced by a blue variety, touted as excellent for digestion. Then strange men in tight-fitting suits appear in the streets. They quietly "eliminate" corrupt local officials, and set up stations where citizens can sell healthy stomach juices for hard cash. Wells's vampire–Martians, who injected themselves with human blood, now become entrepreneurs who, to the profit of some humans, farm others to obtain the product they need: human stomach juice. Narrator Apollo is himself so busy milking the system that he never realizes the Martians are almost literally milking him and his fellow citizens. In a sense, this new regime, with its polite but efficient secret police, has replaced the traditional Russian (and Soviet) black market with a free market, which boosts the economy as farmers begin to profit from blue wheat. In the name of stability (and increased personal gain) Apollo accepts the new dispensation, and takes to drinking blue beer as if nothing has happened.

At the end of *War of the Worlds*, even if humans fail the test, Wells's narrator sees men's views "broadened." As he (probably falsely) sees it, because humans can no longer live in isolation, they are forced to take a speculative, indeed "dialectical" view of the future: "If the Martians can reach Venus, there is no need to suppose that the thing is impossible for men.... It may be, on the other hand, that the destruction of the Martians is only a reprieve. To them, and not to us, perhaps, is the future ordained" (*War* 189). In the Strugatskys' second invasion, there is also a human who notices and learns: Apollo's rebellious son-in-law Charon. This ferryman to the living dead seems a bit more alive than those he ferries. He sees the town's capitulation to Martian economics as stifling what he sees as the basic human need: the need for meaningful work, which is the sole engine for science and progress. Curiously, his description of the Martian takeover reads like a Marxist version of the Don Siegel film *Invasion of the Body Snatchers* (1951). In the latter, "pods" are vampiric invaders from space, who physically take over the human host and rob it of its individual identity; the ensuing "units" are easily shaped into a "classless society," but at the expense of the creative drive that defines the individual human. The Strugatskys' Martians, in opposite manner, impose individuality and laissez-faire economics by duress. Here it is not collectivism but rampant individualism that leads society to inertia. This opiate of material plenty suppresses any need to struggle for a more ideal human world, stifling at the core any need for useful work — work as an end in itself, as the defining element in human existence.

Charon's vision sounds noble perhaps. But if Apollo and his fellows did not have this work ethic after the invasion, they did not have it before the invasion either. Apollo's world is the world Wells's narrator sees before and after the Martian invasion: a world of small shopkeepers who will make any compromise necessary to maintain "stability." Wells's novel, in fact, has its equivalent to Charon — the Artilleryman. This latter is a self-proclaimed "revolutionary" who promotes hard work as the way to human survival but is perfectly willing to let others do the work while he sips wine and smokes fine cigars. His imagined rise to power, driven by class hatred, offers little more than a new redistribution of haves and have-nots. The Strugatskys presented this doctrine of meaningful work in earlier works. For example, Ivan Zhilin, the engineer in *Space Apprentice* who concludes it is better to roll up one's sleeves on Earth than to seek adventure and personal glory in space, returns as protagonist in the novel *Khischchnye veshchi veka* [Predatory Things of our Time, 1965, translated as *The Final Circle of Paradise*], to find himself embroiled in a future world completely given over to gangsterism and terminal ennui — a totally Martianized world. Zhilin discovers that he and this world share a common trait — they are both all too human. Less inconclusively, *The Second Invasion from Mars* ends with the triumph of unregenerate human nature, the lesson painfully learned by Wells's narrator. The glorious future of Marxism turns out, in the light of Martian invasion, to be a reprieve at best.

In two other mature novels, there are alien intrusions. Not only has the emphasis shifted from scenes of destruction to bumbling human reactions to the alien presence, but a new kind of hero emerges. In the 1971 novel *Obitaemyi ostrov* [The Inhabited Island], translated as *Prisoners of Power*, we have another Anton figure, Maxim Kammerer. Here, too, "advanced" socialist humans invade a less advanced world in order to change the course of its history. In the subsequent novel, *Gadkie lebedi* (1972, translated as *Ugly Swans*), we are back to aliens invading us. This time they are not specifically from Mars as in *Second Invasion*. But they are strange beings who suddenly appear in the midst of a provincial, again clearly Russian, town. As seemingly endless rain falls, strange leper-like "rain men" are suddenly everywhere. They engage in philosophical discussions with decadent townsmen, further depressed by the dreary weather. Their "dialectics" seduces the town children and, like the Pied Piper, they eventually lead the children away from their parents toward a new utopian dawning. The novel ends with what seems an apocalyptic moment—sunshine suddenly parts the clouds, and the children, who seem to represent the future, appear clad in white. This time, the alien that captures these children seems to come, not from Mars, but from deep within the old Russian culture itself. The epiphany echoes Tolstoy. In this novel, the invaders, as in *Second Invasion*, seem indistinguishable from the system itself. And yet, this time, it is from inside the system's leper hospitals that the new "future" is born. And it is a future that appears to be a return to what has now become, in the midst of Soviet-style control, a past that seems all the more mythical for being lost and found once again, in the form of the white-clad children that displace the alien.

Even so, the shadow of Wells's novel remains. Children here do not evolve out of their parents—Lenin's spiral of history. Instead they break radically with their past, just as Wells's narrator, after survived the apocalyptic moment of Martian destruction, finds himself in a world where what all that was familiar is now strange. He soon sees, however, that this fantastic twist on things is only a "reprieve" in an endless cycle of violence. And here we must ask just how viable is these children's brave new world. The final scene of the novel presents them, barefoot in white robes, firing stick-guns at a fighter jet that flies ominously overhead. However they mutate, Martians always seem to be among us in the Strugatskys' novels. They invade the amorphous mass of humanity. They force it to extremes, they all but destroy mankind, before the microbe of common humanity destroys them and restores again the human norm—the tourists gawking at the Martian machine on Primrose Hill—that seems doomed to summon them back again.

In this novel, however, the Strugatskys introduce a hero who does not belong to this Martian dynamic that has replaced Marxist history, but rather to Russian folk history. The real focus of this narrative is Victor Banev, a writer and war hero with a record of futile cavalry charges against German tanks, who seeks with all his incorrigible human failings to negotiate this new ideological invasion, against which his writings and his example alike prove ineffectual. The Strugatskys have superimposed over Wells's narrator the classic unruly Russian folk hero, whose struggle is not with history so much as with the contradictory aspects of his own nature. Just as Wells's novel finally comes to rest on the human middle, not on the extremes of Martians or microbes, so here it is neither the rain men nor the children of the new utopian dispensation that attract our interest. It is Victor's blustering and altruistic folly that gains our sympathy. If the children and rain men are the "ugly swans," compounds of paradoxical extremes. Victor is the ugly duckling who holds the center.

A second novel of 1972, *Piknik na obochine* [Roadside Picnic], offers an alien invasion so elusive that it places the entire burden of what is happening on the human inability to understand the nature of the event. A "zone" suddenly appears in the midst of an unidentified (but again quite Russian) place on Earth. The zone harbors a number of inexplicable, highly dangerous, objects. The scientist Pilman offers a tongue-in-cheek surmise: These aliens did not intend to invade or to destroy us, they merely stopped for a picnic, and left behind their litter. For Pilman, any "hostile Martian" theory is bad science fiction. In fact, in a parody of Wells's narrator's list of "benefits" from Martian weapons left behind, Pilman rattles off a list of "leaps forward" in science that can ensue from the objects in the zone: "At least we're using some things — the 'so-sos' and the bracelets to stimulate life processes. And the various types of quasibiological masses, which have created a revolution in medicine" (Roadside, 111). Even so, the focus of the novel is squarely on the inadequacy of human reactions to an alien event. Because we do not understand the function of these things, we misuse them: "I am positive that ... we are hammering nails with microscopes" (Roadside, 111). Even so, misuse is not the whole story. Because of fundamental flaws in human nature, we use these objects for ends of greed, vanity, or power.

Protagonist Red Schuhart, like Victor Banev, is another case of the Strugatskys grafting the Russian folk figure of the crafty survivor upon Wells's narrator. In his actions, Red proves capable, like Banev, of mixing self-interest and utopian vision. In fact, the scene of the Golden Ball at the end of *Roadside Picnic* offers an ironic echo of the final vision of Wells's narrator, in which altruism and hard-headed survivalism collide. Throughout the novel the promise of the Golden Ball has drawn scoundrels, adventurers and idealists alike into the chaos of the zone, each in search of a personal utopia. In the final scene, Red leads Arthur, a handsome idealistic youth, to the ball. Seeing it, Arthur runs recklessly to destruction in the "meat grinder," shouting "HAPPINESS FOR EVERYBODY, FREE ... AND NO ONE WILL GO AWAY UNSATISFIED" (Roadside, 153). Red knows this vision is folly and is a license for social anarchy. Red, however, reflecting on the human disasters wrought by selfish incursions into the zone, also sees a possibility for a collective social consciousness emerging from this alien invasion. Wells's narrator, in the throes of estrangement, still has his wife's hand to hold. Red, likewise, seeks to build on the barest hint of hope. He no longer cynically repeats that human beings cannot, by nature, realize utopia. Instead he says he was never taught to think in a proper utopian manner: "The bastards didn't let me learn how to think" (Roadside, 153). If the statement remains negative, it expresses hope that the human heart might not be inherently evil, but that it might in fact learn to be better. Red's final words, like those of Wells's narrator, offer little consolation. But if they do not eliminate depravity, they relocate its source, not in our "untaught" animal state, but in the social institutions that, failing to teach, have further corrupted human nature. As the Strugatskys' vision matures, it seems increasingly drawn back to the pessimism of *War of the Worlds*, where society and the human heart alike reveal terminal flaws under the extreme pressure of the alien, an alien which (as both Wells's narrator and Pilman suggest) is little more than a mirror of ourselves.

Conclusion: The Invisible Man Comes to Mars

The novel *Za millard let do kontsa sveta* [A Million Years to the End of the World] (1978) translated as *Definitely Maybe*, is the Strugatskys' most complex use of the Wellsian subtext. For the first time in their work, the Strugatskys create a subtle dialectic between two Wells texts — *War of the Worlds* and *The Invisible Man* — by associating the unruly Russian hero (Banev, Schuhart) with the idea of invisibility. This combination offers new possibilities of resistance to a Martian invasion that, in this case, has become a property systemic to the natural world itself. Also the Strugatskys, for the first time, set their novel in a recognizable contemporary locale — Leningrad, which in later works will revert to being the "city of Peter." Finally, in this novel they openly draw on Wellsian sources that are not the censor-approved texts but instead icons taken from the popular Western films derived from Wells's novels: the James Whale and George Pal films respectively.

The novel recounts a new form of Martian invasion, which seems to come this time from physical forces beyond human control. A number of scientists and artists in Leningrad are on the verge of major breakthroughs in their disciplines when mysterious impediments arise: Old mistresses and illegitimate children show up at the door, airline tickets are mysteriously lost, there are sudden deliveries of luxury foods — things that divert them from completing their work. The Strugatskys openly depict a modern Soviet city in the grips of a self-perpetuating bureaucracy. These intellectuals, despite the restraints this bureaucracy has placed on them, have all reached the point of breakthrough, when new happenings, events that appear to be totally random, not traceable to the bureaucracy, suddenly invade and disrupt their lives. The senior scientist among them, however, cannot accept randomness as the cause. He (like Pilman) offers an explanation that elevates these interventions to a new order of reality: He sees these happenings as due to a "homeostatic universe," seen as the tendency of nature itself to defend its equilibrium by acting to prevent human activities that threaten to disrupt that equilibrium. The invasion scenario, in Vecherovsky's theory, has been raised to the level of natural law, against which humans appear totally powerless.

Vecherovsky presents himself as a sort of Kierkegaardian knight of faith, locked in an absurd struggle with the forces of homeostasis, doing battle this time on the level of statistical chaos. The central consciousness of the novel is Malianov, a physicist whose theory of "M-cavities" in some way threatens the general order of things. Malianov does not accept Vecherovsky's pretention to be a knight of faith or a new Luther struggling with the Devil. What he sees instead is a Martian. Vecherovsky's discourse, he says, is accompanied by "satisfied Martian laughter." He goes on to equate this laughter with "the sound H.G. Wells's Martians make when they drank human blood" (Definitely, 49). Clearly, Malianov is not thinking of the novel here, but of the 1953 film version. In the novel, Martians neither laugh nor drink blood. The shift of sources is significant however, for it opposes film and book. In the latter, now associated with Vecherovsky's high culture, Martians have lost their human characteristics, and have become indistinguishable from their machines. Malianov's remark, in a sense, restores a human face to the mechanical entity, even if it is the face of the primitive vampire. Vecherovsky, in fact, at this moment of his proclaimed heroism, is asking each of his colleagues to let him take their life's work with him to the Pamirs, where in isolation he will wage his struggle with homeostasis. Their life's blood is not drunk, it is imbibed in an act of intellectual cannibalism.

If there are, as the title states, a billion years to the end of the Earth, and if all these years are regulated by Martian homeostasis, then how can humanity struggle against what appear to be random invasions? Vecherovsky, it seems, like Anton, has also turned back on the anisotropic road, but this time the road is that of history at the level of natural forces. In turning back, Vecherovsky becomes the victim of his own homeostatic vision of the universe: "I was told the road would take me to the ocean of death, and turned back halfway. Since then, crooked, roundabout, godforsaken paths spread out before me" (Definitely, 104). Malianov, a victim of these intrusions, is nevertheless he who identifies Vecherovsky as a Martian, as one satisfied to accept this dispensation: "And his satisfied guffaws, like Wells's Martian laughter, rang in my ears" (143). Again, however, Wells's Martians don't laugh. Malianov is the opposite of Vecherovsky, always putting a human face on the invader where Vecherovsky (as with Wells's Martians) renders that invader an entity beyond human reach. Wells's narrator would come to dismiss this human face as a failing, yet Malianov realizes that his all-too-human presence can be the means of resisting whatever forces have singled him out. Indeed, he realizes that his actions are so ordinary that he simply loses himself in the crowd, becomes invisible in his ordinary everyday activities. In other words, he becomes the Invisible Man. Reference to Wells's character first occurs when Malianov is visited by a certain Igor Petrovich Zykov of the Criminal Investigation Department, who accuses him of a murder he has no notion of. "The young man was wearing jeans, a black shirt with the sleeves rolled up, and large sunglasses, just like the Tonton Macoute" (Definitely, 27). Zykov does not like the comparison: "Who do you think I look like ... the Invisible Man?" (Definitely, 33). Just as Malianov's Martian comparison is drawn from the film rather than the book, here Zykov's reference is to the Claude Rains character in the film *The Invisible Man* (1933), whose glasses conceal empty eye sockets and garments move without any visible wearer. By mentioning the film version, Zykov associates himself with the destructive side of Wells's figure: he is a Tonton Macoute with no face, capable of anonymous acts of repression and violence. "Invisibility" in his case is the license to commit random acts without taking responsibility.

There are different sorts of invisibility in Wells's work. Griffin is like Vecherovsky, the scientist for whom genius is a solitary act, whose absolute belief in reason is, itself,

Martian. His counterpart Kemp, as upholder of social responsibility, stands for the invisible forces that defend the realm of appearances, of masks instead of faces. Kemp in fact is the prototype of a Zykov, a being who, in serving the commonweal, wields power invisibly, with neither personal commitment nor responsibility. Finally, there is Thomas Marvel, the bumbler who, randomly, ends up with the money Griffin has stolen, and with it purchases the invisibility that allows him to hide the three volumes of Griffin's scientific papers, volumes whose potential impact on human advancement for good or evil is great, but of which he has not the slightest understanding. Malianov, having to face the homeostatic universe, realizes he must assume a like invisibility. As the novel progresses, he comes to see himself as socially invisible: "As long as I obeyed the laws created by the system ... I was protected from all imaginable dangers.... Now, something in the world around me had gone haywire ... I was doomed to perish in front of everyone's eyes" (Definitely, 92–93).

In the end, as Malianov comments on Vecherovsky's "Martian heroism," he stands still clutching to his stomach the white envelope containing his theory of M-cavities. He does not give up his work, but takes it back, unnoticed, to the anonymity of everyday life: "And I will stay at home, meet my mother-in-law and Bobchik at the plane tomorrow, and we'll all go out and buy bookcases together" (Definitely, 142). This is Marvel's strategy exactly: "And on Sunday mornings ... when he is closed to the outer world, he locks the door and examines the blinds.... And then, satisfied with his solitude, he unlocks the cupboard and a box in the cupboard and a drawer in that box, and produces three volumes in brown leather" (Invisible, 125). A billion years lies before Marvel, Malianov, and other invisible men to follow. It is not the narrator of the Martian invasion but the Invisible Man's unheroic double who, in the end, is charged with the task of carrying on science. In the Strugatskys' world, Malianov comes to be the true scientific hero. For he is the one who, under assault from Martians in his society, and now from the Martians who legislate nature as well, is doomed physically to enact Vecherovsky's words, to walk the probabilistic paths of an ever-branching future: "Since then crooked, god-forsaken paths stretched out before me" (Definitely, 142).

In a sense, it is these early SF works of Wells, officially sanctioned but clearly misunderstood by Soviet censors, and especially *War of the Worlds*, that provided the intertextual path whereby the Strugatskys found their way back from the anisotropic road of Marxist historicism, to the "crooked" vision and values of traditional Russian literature. The journey is one from Wells's Mars, a place of unregenerate human nature and its neverending wars of defeat, to the return of the Invisible Man — in this case the long-suffering, all-enduring hero of the Russian folk tale and novel, suppressed by Soviet censors but openly resurrected in figures like Solzhenitsyn's Kostoglotov and the Strugatskys' Malianov.

Notes

1. See Adelaida Lyubimova and Boris Proskurnin, "H.G. Wells in Russian Literary Criticism, 1890s–1940s," p. 63.
2. See Ivan Yefremov, "Heart of the Serpent" in Isaac Asimov, ed. *More Soviet Science Fiction*

(New York: Collier Books, 1962), pp. 19–89. Yefremov conforms to the Soviet ideal, which says that any advanced race must evolve as rational socialist humanity has done. When the crew of the starship *Tellur* makes contact with an alien race, they are our exact mirror image, except for the fact that their metabolism is silicon-based and ours is carbon-based. If love springs up between members of the two races, it must occur through a protective glass wall, for tragically real physical contact would be fatal to both.

3. Stephen W. Potts, *The Second Marxian Invasion: The Fiction of the Strugatsky Brothers* (San Bernardino: Borgo Press, 1991. Potts refers to the "Greek" village of the novel as "an updated version of the settings of much nineteenth-century Russian literature"(61).

Works Cited

Strugatsky, Arkady, and Boris Strugatsky. *Definitely Maybe*. New York: Collier Books, 1978.
_____. *Hard to Be a God*. New York: The Seabury Press, 1973.
_____. *Noon: 22nd Century*. New York: Macmillan, 1978.
_____. *Roadside Picnic*. New York: Macmillan, 1977.
_____. *The Second Invasion from Mars*. New York: Macmillan, 1979.
_____. *Space Apprentice*. New York: Macmillan, 1981.
_____. *The Ugly Swans*. New York: Macmillan, 1979.
Wells, H.G. "The Crystal Egg," in *Twenty-Eight Science Fiction Stories*. New York: Dover, 1952.
_____. *The Invisible Man*. New York: Popular Library, 1964.
_____. *The War of the Worlds*. New York: Berkley Science Fiction, 1964.
Zamyatin, Evgeny. *We*. Trans. Mirra Ginsberg. New York: Avon, 1987.

SAVAGERY ON MARS
Representations of the Primitive in Brackett and Burroughs
Diane Newell and Victoria Lamont

Mars in the first half of the 20th century was the mysterious terrestrial planet next door that, of all the planets in our solar system other than Earth, seemed the most likely to support life; thus, Mars offered a special setting for the early 20th century planetary romances of the sort embraced by Edgar Rice Burroughs (1875–1950) and, later, science fiction writers of classic space opera. Science fiction critics have noted that Burroughs' influence within science fiction rivals that of H.G. Wells (Clute and Pringle 177); of the handful or so of "serious writers" indebted to Burroughs, Leigh Brackett was in the forefront. The others in that category are Ray Bradbury, Michael Moorcock, and "above all" Philip José Farmer (178). From a young age Brackett (1915–1978) was an avid reader of Burroughs' Mars and Venus stories, starting with *The Gods of Mars* (1918) (Hamilton vii; Truesdale 9): "The Mars stories, all my Mars stories came out of Burroughs ... my fascination for Mars came from the fascination for *his* Mars," she once recalled (Truesdale 10). Through her mid-century Mars and Venus space opera that appeared in the pulpier magazines, notably *Planet Stories*, Brackett became a master creator of moods and the undisputed "queen" of space opera, fashioning a relatively dark form of classic space opera and contributing to the development of science fantasy (Stableford, "Creators" 41, and see Newell and Lamont). However, there has been little in-depth analysis of the intertextual relationship between the two authors. Such analysis is enlightening, for Brackett was not a mere imitator of Burroughs but an innovator in her own right. Michael Moorcock in his introduction to the Brackett collection *Martian Quest: The Early Brackett* (xiii) even suggests that the best of Brackett's Martian adventure stories actually surpass those of Burroughs.

Burroughs' Barsoom (Mars) series, which began with *A Princess of Mars* (1912), and was extended to eleven volumes over thirty years, was written in the tradition of

American frontier narratives that were extremely popular when Burroughs began writing stories for the pulps in 1912. Burroughs' hero of the earliest of them, John Carter, was himself a southern gentleman exploring Arizona when he was mysteriously transported to Mars. Whereas Burroughs was careful always to preserve his hero's status as a true country gentleman and heroic representative of civilized values on the barbaric Martian frontier, Brackett's Eric John Stark, first created for the science fiction and fantasy pulps in 1949 and reworked and reintroduced in the 1970s, was a much more ambiguous figure. Born on a mining colony on Mercury, Stark is orphaned when the colony is destroyed, and is raised by the planet's indigenous and savage inhabitants. He is later taken prisoner by subsequent human colonizers, who display him like a zoo animal until he is finally rescued and adopted by Simon Ashton, a representative of the "Earth Ministry" who "civilizes" Stark and to whom Stark remains fiercely loyal throughout Brackett's many novellas about him. With his allegiances thus divided between the indigenous peoples of the various planets targeted for colonization, and Simon Ashton's participation in efforts to "civilize" the galaxy, Stark embodies an ambiguous attitude toward colonization that is especially pronounced in Brackett's early Mars stories, and in her fantasy Venus and fantasy Mercury stories that Brian Stableford suggests are "simple extensions" of Brackett's Martian mythology ("Best" 26). Writing during the immediate post–World War II period of America's increasing international presence, Brackett took an attitude toward the spread of civilization far more nuanced than that of Burroughs, who took for granted the imperialist values that are as pervasive in his Mars stories as they were in early 20th century American culture.

As a writer for the pulps, Brackett was under pressure to write quickly and for profit, and unlike Burroughs, with Brackett there was no series continuity joining her Martian settings (Clute 150), yet she developed a complex diegetic universe across the many pulp stories she wrote about Mars during the 1940s and 1950s. As Brian Stableford reminds us, at the very time that the venerable editor of *Astounding Science-Fiction*, John W. Campbell, Jr., was successfully promoting a more "scientific and realistic" image of the planet, Brackett (and her friend Ray Bradbury) headed in the opposite direction, creating "a calculatedly nostalgic version of Burroughsian Mars" ("Creators" 41). Brackett's original Mars was, writes Stableford, a "decadent ancient planet facing the threat of plundering Earthmen" ("Mars" 778). Decades later, Brackett would write that for her, despite what planetary science has taught us about actual Mars, "Mars is still fun" as a creation of a writer's imagination, full of wonders ("Introduction" 5).

Like Burroughs writing about a fictional Mars in the 1910s, Brackett in the 1940s and 1950s drew from American frontier mythology to represent Mars as a far more ancient planet than Earth, a planet whose "civilized" period was long past and was now in a state of atavism. But while the Mars stories of both authors can be read as narratives of American empire, their ideological stances are quite different. Burroughs' first Barsoom series novel, *A Princess of Mars*, closely follows the formula of popular historical novels in the United States at the turn of the last century, which Amy Kaplan has persuasively argued had more to do with American imperial incursions into Mexico and the Philippines than with the more exotic settings of texts such as Richard Harding

Davis' *Soldiers of Fortune* (1897, set in present-day Latin America) and Charles Majors' *When Knighthood Was in Flower* (1898, set in Tudor England). In countless novels of this type, the story begins with the hero languishing in the tedium of established civilization, until the opportunity to involve himself in some colonial exploit, usually involving Civil War in a foreign land and the rescue of the princess of the more righteous of the warring tribes, regenerates his under-tested masculine virtues. Burroughs' plot follows this pattern almost exactly: His hero, John Carter, is a post–Civil War southern gentleman and army captain who, upon the close of the Civil War, seeks his fortune in the gold mines of Arizona. After stumbling upon a mysterious cave during an altercation with hostile Apaches, he is mysteriously transported to Mars (called Barsoom by its indigenous inhabitants). The lesser gravity of Mars gives him superhuman powers — transforms him into a "superman," according to Brian Attebery (63) — that enable him to quickly progress up the social ladder of a race of monstrous, war-mongering, green people. Like his counterparts in Earthbound historical romances of the period, Carter engages in military heroics that end Civil War on Mars and win him a Martian name (Djor Sojat) and the hand of the Martian Princess of Helium, Dejah Thoris. Also as in historical romances of the period, Carter is fated not to remain on the colony indefinitely but, in keeping with myths of American Empire of the period, must leave the rule of the colony to its native inhabitants in order to sustain the mythology of the benevolent and democratic goals of American Empire. In the other two Barsoom series novels featuring John Carter, *The Gods of Mars* and *The Warlord of Mars* (1919), the hero revisits his Martian family for similar adventures.

While Brackett's Eric John Stark is clearly part of the tradition of the planetary romance, which is in turn, as we have shown, in the tradition of the American historical romance, her representation of colonialism and of her hero's relation to it in the 1940s and 1950s is far different from that of Burroughs. First appearing in *Planet Stories* in 1949, Brackett's Eric John Stark is a forerunner to the morally ambivalent frontier heroes of film westerns of the 1960s, particularly the Man With No Name, played by Clint Eastwood in Sergio Leone's trilogy *A Fistful of Dollars* (1964), *For a Few Dollars More* (1965), and *The Good, the Bad, and the Ugly* (1966). Whereas John Carter embodies unambiguously American cultural superiority over the declining civilization of Barsoom, Stark is introduced as a mercenary with ambiguous allegiances. With his allegiances divided between his savage adopted tribe and his civilizing foster-father Simon Ashton, who works for an agency implicated in the administration of Earth's interplanetary empire, Stark, at least in his first incarnation in the novellas Brackett wrote for *Planet Stories* in the early 1950s, is as likely to take the side of the indigenous inhabitants of Mars and Venus as he is that of the colonizers. In "Queen of the Martian Catacombs" (1949), for example, Stark exposes the plot of a Martian chieftain to install himself in a puppet government propped up by colonial "outlanders." In "The Enchantress of Venus" (1949) Stark searches for a friend who has disappeared and finds himself entangled in the attempts of the declining ruling class to maintain their tyrannical hold over the people of Shuruun. In both cases, Stark is offered the opportunity to rule at the side of a conniving princess if he will only help prop up her rule, and he refuses. This is a point-for-point reversal of the adventure romance plot from which the planetary romance was derived, in which the hero typically wins the affections of the indigenous princess and rules, albeit temporarily, at her side. In early Stark stories, the hero is more likely to sympathize with the point of view of ordinary subjects than with their rulers, and is often at odds with the goals of Earth's interplanetary empire.

Brackett's engagement with the concept of race also sets her apart from Burroughs in interesting ways. In their representation of race, gender, and colonization, both authors enter into an American literary tradition that can be traced back at least to the 1820s with frontier narratives such as James Fenimore Cooper's *Last of the Mohicans* (1826) and Lydia Maria Child's *Hobomok* (1824). In these tales of colonial race relations, attractions between white and mixed-race characters figure colonial political dynamics: In *Last of the Mohicans*, for example, the mixed-race heroine Cora is doomed to a tragic fate because she is out of bounds for potential white suitors, while her white blood elevates her above other indigenous women and makes her the target of the evil and lustful Indian Magua. In this way, Cora embodies the future — or rather lack thereof— of a racially mixed American body politic. Like Cooper's representation of the Mohicans, Burroughsian Mars is a civilization in decline, a dying world. When Carter first arrives there he encounters the vestiges of a once-great civilization, including grand cities, monumental buildings, and great artwork. But, Carter gradually learns from Princess Dejah, the ancient, white-race creators of this civilization are extinct, having joined with other great races of early Martians "who were dark, almost black, and also with the reddish yellow race that had flourished at the same time" (*Princess* Chapter 11) to form Princess Dejah's hybrid race, which is now at war both internally and with the green people. Carter, meanwhile, represents a race in its prime, naturally fitted to help restore this declining civilization to its former peace and prosperity. In Burroughs' first Barsoom novel, Carter's exploits on Mars culminate in his marriage to Dejah, making him the father of a future race of rulers on Mars. Mars is thus regenerated by the blood of the racially pure John Carter.

In contrast to Burroughs' straightforward adaptation of the ideology of Anglo-Saxon racial superiority to an intergalactic setting, Brackett's projection of race and the primitive is far less reinforcing of the standard privileged categories of colonial discourse. Although Stark's biological parents were Terran (meaning "from Earth"), Stark is figured as a racially mixed character whose skin has been burnt "dark" by the hot sun of his home planet, Mercury. Whereas John Carter's presence on the Martian colony purifies and regenerates it, Stark's very body is permanently marked by the colonial environment of his upbringing. Even though Stark is "civilized" by his savior Simon Ashton, he is still haunted by his primitive upbringing and has frequent flashbacks to his childhood when he bore the name N'Chaka and lived in a constant state of vigilance. Civilization has not completely supplanted the primitive part of Stark's character, which surfaces during moments of threat and indeed is the source of his superiority over his enemies. Stark frequently feels fear, a sensation linked to his primitivity — but one which gives him an advantage because it enables him to respond to threat with the instinctive quickness of an agile wild animal. This is a departure from the frontier romance tradition, which treats emotion as primitive, racialized, and, in white men, subject to the mastery of reason and will. It is Stark's *inability* to master his fear that enables him to defeat his enemies. Stark's racial hybridity enables him to serve as an instrument of the colonial center because he can travel more freely among different peoples and planets; therefore, he is often recruited by Simon Ashton for missions that serve the interests of the empire in some way. However, Stark is not modeled in the romantic tradition in the same way that John Carter is: whereas the romantic hero enters into a marriage that resolves conflict between warring factions and produces an idealized ending, Stark's hybrid status prevents him from entering into more than temporary alliances, and he challenges colonial authority as often as he serves it. In *People of the Talisman* (1964), for example, his dying Martian friend entrusts Stark with the care of a precious talisman rumored to have the power to bring a great Martian leader back from the dead.

Among Brackett's particularly inventive details of Martian "local color" are her narratives of the local practice of "Shanga." Participants in this ritual are exposed to a form of radiation that causes "artificial atavism," allowing men to temporarily "wallow in beasthood" ("Queen" 17). Whereas John Carter is able, à la Cooper's Leatherstocking, to live among the primitives without ever becoming one of them, Brackett represents the primitive as the object of colonial desire in her 1948 novelette "The Beast-Jewel of Mars." In this story, Shanga is an outlawed, underground practice particularly popular with colonists seeking relief from "the jangled nerves and overwrought emotions of modern man" (6). Whereas John Carter criticizes "the loss of all the finer feelings and higher humanitarian instincts among [the] poor creatures" of Mars (*Princess* 67), Brackett's colonists actively seek the release from inhibition and indulgence in primitive drives that Shanga gives them, and are indeed more susceptible to its influence than are the indigenous inhabitants.

Brackett introduced Eric John Stark in some of her earliest stories of the 1940s and 1950s. However, there was no series continuity joining her Stark stories, so she did not become known for that character until the early 1970s, when she published a trilogy of novels featuring an updated version of the Stark character: *The Ginger Star* (1974), *The Hounds of Skaith* (1974), and *The Reavers of Skaith* (1976). This was a marketing strategy that shifted the focus of Brackett's "brand" away from her favored settings of Mars and Venus—which space exploration had to a certain degree rendered unusable as a setting for planetary romance—and onto the heroic figure she created to explore them. It was these later novels that established Brackett's reputation as the creator of Eric John Stark, effectively supplanting her previous accomplishments in the science fiction field. However, these later Stark novels were quite different from her earlier stories, representing colonial contact as liberating rather than exploitative. They depict Stark's adventures on the distant, fictional planet of Skaith, whose sun is dying. Skaith has only recently come into contact with the intergalactic community, a development which threatens the tight grip of its repressive leaders over the population. Stark leads a crusade to open up the planet to intergalactic travel so that the people of Skaith have the opportunity to emigrate and thus survive. While Brackett still aligns her hero with the inhabitants of the colony, she presents colonialism as a liberating rather than an exploitative process—but it is liberating for different reasons than those envisioned by Burroughs.

Whereas Burroughs' Mars is racially regenerated by the colonial hero, the people of Brackett's Skaith can survive only by leaving their homeland behind and joining the intergalactic community. Brackett's updated vision of colonialism is in keeping with her support for the American involvement in Vietnam in the late 1960s and early 1970s and, more generally, with the emergence of the neo-colonialism of the 1970s, '80s, and '90s, in which participation in global markets and communities was increasingly touted as the solution to localized economic pressures and injustices.

For both Brackett and Burroughs, Mars functioned as a bridge between the known and real and the unknown and fantastic. Its relative proximity to Earth and the tantalizing prospect — at least until the discoveries touched off by NASA's *Mariner* and *Viking* space missions in the mid–1960s and mid–1970s, respectively — that it could be habitable, thus a future home for Earthlings (the surface features of Mars were visible using optical telescopes, leading 19th century scientists and fiction writers to speculate about the planet's capacity for supporting life outside the Earth; see Stableford, "Mars.") provided Burroughs and his immediate predecessors with a usable setting for colonial narrative at a time when political realities on Earth were testing the viability of such narratives. In 2010, the jury is still out on this score. The American frontier "safety valve" was not unlimited, and when it ran out, on what basis could America continue to claim exceptional status as a frontiering nation? Burroughs turned to Mars to help sustain a little longer the mythology of a nation predicated on expansion into alien territory. He chose Mars as his setting because, as fantastic a setting as it was, it was a *real* place and therefore a potential territory into which the empire could one day expand. Brackett too exploited this liminal aspect of Mars as a setting both real and fantastic, moving against the trend in science fiction and the mounting scientific skepticism of the immediate postwar era that Mars was capable of supporting human life, but she complicated colonial ideology by refusing to allow romantic adventure to gloss over its economic underpinnings. Her Mars of the 1940s and 1950s is the exploited victim of Earth's insatiable desire for power, territory, and resources as much as it is a proving ground for colonial heroics. But by the 1970s, the reality of Mars made even Brackett's more cynical vision seem too romantic, for the reality of the day proved to be that of a barren and lifeless world. While she found Mars to be still fun and predicted it would be back as a topic and setting for stories ("Introduction" 5), Brackett did not live long enough to find out. Mars did come back in the fiction of excellent SF writers such as Kim Stanley Robinson, who employed Mars as a setting his acclaimed Mars trilogy on the themes of ecology and sociology. Henceforth, narratives of intergalactic empire became more and more the stuff of pure fantasy as humans were confronted by the reality of limited resources on their small and lonely planet.

Acknowledgments

The authors wish to thank the organizers of the 2008 J. Lloyd Eaton Conference on the theme "Chronicling Mars," and in particular, George Slusser, for the opportunity to present a shorter version of this paper. They are grateful for the assistance of the staff of the Merril Collection of

Science Fiction, Speculation, and Fantasy, Toronto, and for the financial support of the Peter Wall Institute of Advanced Studies, University of British Columbia.

Works Cited

Atteberry, Brian. *Decoding Gender in Science Fiction.* New York: Routledge, 2002.
Brackett, Leigh. "The Beast-Jewel of Mars." *Planet Stories,* Winter 1948.
_____. "Black Amazon of Mars." *Planet Stories,* March 1951.
_____. "Enchantress of Venus." *Planet Stories,* Fall 1949.
_____. *The Ginger Star* [Reintroducing Eric John Stark]. New York: Ballantine, 1974. First published as a two-part serial in *Worlds If,* February and April 1974.
_____. *The Hounds of Skaith* [Further Adventures of Eric John Stark]. New York: Ballantine, 1974.
_____. "Introduction: Beyond Our Narrow Skies." *The Best of Planet Stories No. 1.* New York: Ballantine, 1975.
_____. "Martian Quest," *Astounding Science Fiction* February 1940.
_____. *People of the Talisman.* New York: Ace, 1964.
_____. "Queen of the Martian Catacombs." *Planet Stories,* Summer 1949.
_____. *The Reavers of Skaith* [Eric John Stark #3]. New York: Ballantine, 1976.
Burroughs, Edgar Rice. *The Gods of Mars.* New York: A.C. McClurg, 1918. From a five-part serial published in 1913.
_____. *A Princess of Mars.* New York: Grosset and Dunlap, 1917. From serialized stories published in 1912.
_____. *The Warlord of Mars.* New York: A.C. McClurg, 1918. From a four-part serial published during the period 1913–1914.
Child, Lydia Maria. *Hobomok.* 1824. Reprint, ed. Carolyn Karcher. New Brunswick, NJ: Rutgers, 1986.
Clute, John. "Leigh Brackett." *The Encyclopedia of Science Fiction.* Ed. John Clute and Peter Nicholls. London: Orbit, 1993.
Clute, John, and David Pringle. "Edgar Rice Burroughs." *The Encyclopedia of Science Fiction.* Ed. John Clute and Peter Nicholls. London: Orbit, 1993.
Cooper, James Fenimore. *Last of the Mohicans.* 1926. Reprint, New York: Penguin, 1986.
Davis, Richard Harding. *Soldiers of Fortune.* New York: Scribner's, 1897.
Hamilton, Edmond. "Story-Teller of Many Worlds." *The Best of Leigh Brackett.* Ed. Edmond Hamilton. New York: Ballantine, 1977.
Kaplan, Amy. "Romancing the Empire." *American Literary History* 2.4 (January 1990).
Majors, Charles. *When Knighthood Was in Flower.* Toronto: G.J. McLeod, 1898.
Moorcock, Michael. "Queen of the Martian Mysteries: An Appreciation of Leigh Brackett." *Martian Quest: The Early Brackett.* Royal Oak, MI: Haffner, 2002.
Newell, Dianne, and Victoria Lamont. "Leigh [Douglas] Brackett, 1915–1978." *Fifty Key Figures in Science Fiction.* Ed. Mark Bould, Andrew M. Butler, Adam Roberts, and Sherryl Vint. New York: Routledge, 2010.
Stableford, Brian. "The Best of Hamilton and Brackett." *Vector* 90 (November–December 1978).
_____. "The Creators of Science Fiction — 5: Leigh Brackett." *Interzone* 104 (February 1996).
_____. "Mars." *The Encyclopedia of Science Fiction.* Ed. John Clute and Peter Nicholls. London: Orbit, 1993.
Truesdale, David, with Paul McGuire. "An Interview with Leigh Brackett and Edmond Hamilton." *Science Fiction Review* 6.2 (1977).

THE (IN)SIGNIFICANCE OF MARS IN THE 1930S

John W. Huntington

For a long time scholars have observed the way *Out of the Silent Planet* (1938) attempts to undercut the "materialism" Lewis objected to in Stapledon's *Last and First Men* (1930). I want to push this difference further, by inserting Wells' *Star Begotten* (1937) into the discussion, by tracing a political-religious difference expressed in the very form of the fictions, and finally by seeing the conversation as not about politics in the immediate sense but about the possibilities and limits of science fiction in a time of political crisis. The differences between Lewis and the two earlier writers are clear and at least on Lewis's part explicit. My point is that, given their basic differences, we can learn something about how the possibilities of SF were thought of at this historical moment. The conservative Lewis pushes SF back toward earlier patterns of the scientific romance, while Stapledon shows in his story and Wells in the very form of his narrative frustration with the romance form, a pessimism and a skepticism that finds that form too neat. The difference is rendered clearly by the way they go about imagining Mars.

When I was editing *Star Begotten* a couple of years ago it didn't occur to me to pursue the significance of the fact that the cosmic rays that cause all the commotion are supposed to come from Mars. I took it for granted that Wells himself did not have any special thoughts about Mars at the time, beyond the obvious and pointed allusion to his own anti-imperial allegory of forty years earlier. I have now come to think about that moment of the 1930s as a remarkable one in which Mars became a site for argument about what we can know — which becomes, because of the genre that is at that time in a crucial moment of formation, what can SF itself think?

In the opening lines of *Star Begotten*, Wells tells us "This is the story of an idea and how it played about in the minds of a number of intelligent people" (37). The novel tells how the "idea" that Martians are making humans into "Martians" by bombarding Earth with mutation-causing cosmic rays gets its start in the chat of a London club, comes to obsess Joseph Davis, an author of popular nationalist histories who is about to become a father, takes over the public press for a season, and then dies out, leaving Davis still believing that he, his infant son, and his wife may be Martians. In my introduction to the edition I argued on the basis of interior evidence in Wells's novel and on the scientific implausibility of the main idea that the "Martian influence" is imagined and that the novel is in part about scientific delusion.

Star Begotten begins by making fun of Joseph Davis as a man of intellectual compromises, but by the end of the story Davis has achieved an enlightenment that inspires him to destroy the manuscript he is writing and to let his previous books go out of copyright. But it is still ambiguous whether this new awareness may not be simply a mutation of the original self-justifying narrative. Perhaps Joseph Davis has not become enlightened but simply reoriented his devotion. The novel's criticism is finally, not directed just at patriotic Anglophilia; it is directed at devotion itself. Davis catches a glimpse of this quality in the schoolboy he discovers in the midst of his "research" who asks, "What is spiritual?"—a question neither Davis nor the school's headmaster can answer (92–94). We never hear of the boy again in the novel, but he has established a standard of skepticism that Joseph Davis even as he reorients himself does not attain. Wells' novel asks us to partake of that skepticism.

There are two moments in *Star Begotten* when Wells pointedly positions the narrative in relation to other pieces of SF. Despite discouraging comments from all sides, the idea of mutating cosmic rays spawns fantasies of "scientific" possibilities which then are accepted as givens. The anonymous "rufous man" in the club suddenly declares, "Suppose these cosmic rays come from Mars!" When a skeptical professor observes that "they come from every direction," the rufous man is undaunted and keeps speculating:

> Including Mars. Yes, Mars, that wizened elder brother of the planet Earth. Mars, where intelligent life has gone on far beyond anything this planet has ever known. Mars, the planet which is being frozen out, exhausted, done for. Some of you may have read a book called *The War of the Worlds*—I forget who wrote it—Jules Verne, Conan Doyle, one of those fellows. But it told how the Martians invaded the world, wanted to colonize it, and exterminate mankind. Hopeless attempt! They couldn't stand the different atmospheric pressure, they couldn't stand the difference in gravitation; bacteria finished them up. Hopeless from the start. The only impossible thing in the story was to imagine that the Martians would be fools enough to try anything of the sort. But—... [s]uppose they say up there, "Let's start varying and modifying life on Earth" [62].

We need to think about what happens in this moment. Wells mocks his own earlier fiction for being scientifically implausible, and yet that debunked fiction morphs into the narrative assumption that generates the subject of the novel. *Star Begotten* is a fiction that repeatedly embeds within itself reminders of how implausible it is and how willfully fanciful those who believe it are.

Sometime later in *Star Begotten* Wells alerts us even to the delusions that speculation under these conditions can arouse. In the middle of the novel Davis and two friends, Professor Keppel and Doctor Holdman Stedding, indulge in some utopian "play" with the idea:

> "One's imagination wants to play with it. It's as attractive as a hare's foot to a kitten. Suppose Keppel, suppose—for the sake of a talk—there are Martians.

"Let's suppose it. I'm more than willing."
"What sort of minds would they have and what would they think of our minds and what might they not try to make of them?"
"Regarded as an exercise in speculative general psychology? That's attractive."
"As a speculative exercise then."
"Exactly. You know that man Olaf Stapledon has already tried something of the sort in a book called *Last and First Men*" [79].

The "speculative exercise" of imagining Martians at this point in *Star Begotten* leads Keppel to a fairly optimistic, familiarly Wellsian, utopian dream of a rational and ethical civilization. But the pointed reference to Stapledon might make us worry about the seductions of that argument. Yes, Stapledon had imagined interplanetary "general psychology," but that exercise had hardly led to a utopian vision. *Last and First Men* is particularly pessimistic in its depiction of the Martians' invasion of Earth under the Second Men. Stapledon's fanatical Martians fail to understand humankind despite over three hundred thousand years of invasion and occupation of Earth, and humankind never grasps the "psychology" of the Martian clouds who, sensitive to electromagnetic radiation, recurrently attack radio stations because they imagine the stations control the Earth. The Martians, being almost immaterial, admire matter, honor diamonds above all, and devote their energies on Earth to the bizarre project of rescuing diamonds from darkness and obscurity and placing them on mountain tops where they belong. As time goes on, the Martians on Earth begin to question the project, especially as they begin to suspect that they are inflicting pain on the previously unrecognized human population, but the Martians back on Mars, fundamentalists of a sort, criticize the colonists for backsliding, and the allegory becomes one of party loyalty overwhelming the actual facts of the situation. The Second Men themselves, despite their considerable mental accomplishments, find themselves caught between a grand, stoic acceptance of reality ("Thus is the world. Seeing the depth we shall see also the height; and we shall praise both" [131, 136]) and a terrifying suspicion that "the human race had utterly deceived itself, and the course of cosmic events after all was not significant, but a meaningless rigmarole" (136). The final outcome of the era of Martian invasion is a plague, created by humans, which wipes out all the Martians and most of the humans. Perhaps even more devastatingly ironic is the fact that the human destruction of the Martians, far from leaving the humans triumphant, causes the Second Men to fall into a deep, race-destroying existential despair.

Keppel's imaginings are much more sympathetic to the Martians. He reasons that they must be more technologically advanced than humans and therefore more ethically advanced. But the novel itself never concedes any validity to such a conclusion. If Keppel has a hope of cross-planetary understanding, his allusion to Stapledon suggests something quite different: a darkly ironic, almost absolute incomprehension. The Martian episode in *Last and First Men* can be taken as a parable of the problem of alienness, such that the Martians never grasp that humans are conscious, and the humans never understand what motivates the Martians to "rescue" diamonds and destroy the sources of radio emissions. Stapledon's imagined Martian invasion hardly encourages utopian speculation. If anything, it warns of the limits of understanding under these conditions.

Keppel's line about the attractiveness of speculation in "alien psychology" may also be a clue to a further dialectical situation wherein it is the very satisfactions of the generic project that should make us suspect self-indulgence. We are here at a crucial distinction between science and fiction. It doesn't take a Popperian to argue that utopian speculation is not what science is about. It is fiction that indulges this pleasure. Science, importantly, requires empirical testing. Stapledon's deluded Martians, while they may claim to be scientific, are victims of the Martian chauvinism that by linking intelligence to radiation has caused them to deduce that humans are cattle and radio transmitters rule the Earth. The comic parable is more than just an ironic glance at "science," however; it is a cautionary story about the pleasures of speculation, about the delusory quality of SF itself.

The one dimension in which Joseph Davis can be seen to grow is in overcoming his spontaneous presumption that other intelligent creatures must also be ethically "lower" than humans. At first, "For some very deep-seated reason in his make-up, it was an intolerable thought for him, that there should appear any class of creatures on Earth intellectually above his own, unless they were profoundly inferior to him morally, and so repulsive and ugly as practically to reverse the handicap against him" (87). This is a line of thought that Lewis claims Wells has taught us. In *Star Begotten*, however, this xenophobia is explicitly criticized, and Joseph Davis later comes to spend time and energy trying to envision the Martians as wise and serene (91). It is important to acknowledge this rejection of a kind of knee-jerk xenophobia, but we should also be aware that neither Joseph Davis nor the novel itself are ever able to depict benign aliens. It never actually imagines the Martians as such. In fact the only "Martians" we ever see are the fairly ordinary human beings Davis imagines to be Martians. Keppel's utopian speculations, though they begin with assumptions about Martians, depend less and less on alien creatures, and by the close of the novel he is talking not about Martians at all but simply future people. Only Joseph Davis continues to harp on Martians, and in seeing himself as already Martian at the end — that is, by accepting the SF idea — he solves the problem of the alien by making it identical to the human, though still, strangely, with an implication that the human has changed.

Lewis' *Out of the Silent Planet* is constantly conscious of Wells. Lewis prefaces his novel (published the year after *Star Begotten*) with a disavowal which amounts to a warning: "NOTE: Certain slighting references to earlier stories of this type which will be found in the following pages have been put there for purely dramatic purposes. The author would be sorry if any reader supposed he was too stupid to have enjoyed Mr. H.G. Wells's fantasies or too ungrateful to acknowledge his debt to them" (6). Some of the "slighting references" are repeated allegations of xenophobia that the novel attributes to Wells. While a generous acknowledgment of debt, the note is also a pointed harking back to an earlier literary idea. Both the genuine appreciation of Wells as the master of the modern scientific romance and the charge that Wells sponsors xenophobia signal how much Lewis' novel is a vigorous attempt to restore the kind of speculation that the rufous man in *Star Begotten* has termed "hopeless." In direct contradiction to *Star Begotten*, which repeatedly and on many layers prevents our finding an unambiguous reading of what Mars represents, Lewis' primitive Martian paradise, depicted in extravagant and startling detail, represents a secure ideological position.

As if in rebuttal to the epistemological questions we see in Stapledon and late Wells about what we can know or understand, Lewis creates a fiction which is brilliantly unambiguous about the facts. It is telling that on Malacandra the confusions felt on first meeting new species fairly quickly resolve into a clear taxonomy. The term "hnau" identifies the category of creatures which "matter." Hrossa, Seroni, Pflifftriggi, and humans are all "hnau," while the hnakra — the only "animal" on Mars, apparently — and eldila — spiritual, angelic creatures — are not "hnau." The curious thing is that no systematic definition of the category "hnau" is ever attempted. Some hnau are identified, but what makes them hnau is never explained. The effect is an illusion of clarity and understanding on an issue of central importance without any actual explanation. The spirit of *Out of the Silent Planet* is to adumbrate a clear and orderly reality by approximations garnered from language study, literary history and, most centrally, Christianity.

The novel positions us to watch and experience Ransom as he endures mystery in order to arrive at understanding and wisdom. Lewis' narrator, hardly noticeable until the final pages of the novel, allows us to entertain the idea that the whole story is just a dream, but this very traditional narrative device should not confuse us about the novel's values. Stapledon's plot, narrated by one of the Last Men who already knows the whole story, might seem to offer a similar narrative situation, but though *Last and First Men* hides nothing, even in its omniscience it can find no solution for the difficulty of alien incomprehension. The Last Man who narrates the history of humanity to the present does not understand the meaning of his impending extinction. If Lewis's subject is the revelation of the mysterious truth behind the dark glass, Stapledon's is not about the true reality so much as it is about the difficulty, perhaps given who we (and the Martians) are, the impossibility, of perceiving the reality. The Martian episode ends with a disabling despair. Wells's enigmatic novel goes further and gives formal shape to the inherent difficulty of knowing what to make of Mars and poses an essentially comic structure where, amidst scenes of delusion, misconstruction, propagandistic excess, and playful, utopian fantasy, we cannot tell for sure what is actually happening or what it means.

At one level, this is the thesis Mark Hillegas posed forty years ago (133–144). But if we think of it as more than just a philosophical argument and remind ourselves of the international situation at the time, we can see how the different novels conceive of SF itself as a source of wisdom in a time of excruciating conflict. Lewis, appealing to the great tradition of Christian fantasy that includes Dante and Spenser, finds in SF an opportunity to write an allegorical parable. Wells and Stapledon, much less confident about the true nature of reality, use SF to present provocative and finally irresolvable narratives that, rather than playing the orthodox generic game, call the very rules of the game into question. In 1937 Wells' Martian novel alerts us to the effects of propaganda and the powers of rumor and panic, and by dignifying skepticism places us in a situation similar to that of Davis himself, unsure of what we can know. Lewis responds with a novel depicting Mars, the warlike planet, as a peaceful utopia. For him, too, it may be a dark time on Earth, but the traditional structures of understanding are secure.

Works Cited

Hillegas, Mark. *The Future as Nightmare: H.G. Wells and the Anti-utopians.* Carbondale: Southern Illinois University Press, 1967.
Lewis, C.S. *Out of the Silent Planet.* New York: Macmillan, 1965.
Popper, Karl R. *The Logic of Scientific Discovery.* New York: Harper Torchbooks, 1965.
Stapledon, Olaf. *Last and First Men.* London: Methuen, 1930.
Wells, H.G. *Star Begotten*, ed. John Huntington. Middletown, CT: Wesleyan University Press, 2006.

Spawn of "Micromégas"
Views of Mars in 1950s France
Bradford Lyau

In 1752 Voltaire published his classic short tale "Micromégas." He tells the story of two giant visitors from other worlds, the bigger one from a planet revolving around Sirius and the other from Saturn, who visit Earth.

These visitors are not only enormous in size, but are also endowed with senses and intelligence far beyond any comprehension by the mere humans who inhabit planet Earth. Of course, these beings are used as props to emphasize that humanity is not the center of creation and that the new scales of the cosmos as discovered by the latest science must force humans to reevaluate how they view themselves in relation to the universe around them.

When looking at the Anticipation line of science fiction novels, published by Fleuve Noir, France's first producer of predominantly native science fiction after World War II, and their portrayal of Earth's outward neighboring planet, Mars, the use of the Red Planet as a device with which humans can view themselves from a different perspective is very evident and can be said to belong firmly in the tradition of "Micromégas."

First of all, Voltaire's story is a tale of scales and proportion, dealing with the new vistas revealed by science, both big and small. If the findings of Galileo and Newton exposed how big the universe really is, the discoveries of people such as Leeuwenhoek reminded many of what remains to be discovered in the infinitesimally small that surrounded all in daily life. How should the uniqueness and dignity of humanity now be viewed? (Wade 61–77).

Many of the Anticipation novels also deal with scales and proportion, but this time the focus is on the new powers of science and technology. How much, or how little, should these two potent activities be applied to social policy, especially during the French government's modernization policies of the 1950s? The Martian societies portrayed in these novels examine science and technology in this light.

Secondly, "Micromégas" is an example of one of Voltaire's noted contributions to literary development, the combining of neoclassical fictional forms with nonfictional efforts in metaphysics, natural philosophy, and moral philosophy to produce the *conte philosophique*, or philosophical tale (Wade 79–88).

The Anticipation stories, meanwhile, could be said to be a combination of two literary traditions, that of the *conte philosophique* tradition, enriched by later developments in French literature, and that of the importation after World War II of American genre forms, including science fiction. (The French did produce their own genre literature, but not in the same manner as in America, and there was a hiatus of publication of science fiction stories in France from the mid–1930s to the end of the Second World War.)

Third and finally, as Ira Wade concludes in his study of "Micromégas," Voltaire's story is one of limitations. Even the two visiting aliens, despite their seeming omnipotence to human eyes, have their limits and must acknowledge them (Wade 115).

The Anticipation novels also explore limitations, this time those of modern science and technology.

The writers to be examined here are F. Richard-Bessière, B.R. Bruss, Kemmel, Maurice Limat, Jimmy Guieu, and Gerard Klein (writing as Gilles d'Argyre). The first two portray Mars as a warning of what could go wrong, the next three use Mars as a savior or guardian of the human race, and the last uses Mars as a backdrop for criticism on policy — or maybe for something very different.

Richard-Bessière (a pseudonym for two writers publishing as one), besides being the first author published in the series, also provides the paradigmatic presentation of this theme of the role of science and technology in society. His first novel, *Les conquerants de l'univers*, describes a visit to Mars where there exists an advanced Martian global society governed by ideas of science and technology.

The chief characteristic of Mars's society is its hierarchical structure, whereby a person's position is determined by the amount of knowledge he or she possesses. The Martians organize themselves into four levels: (1) scientists (*savants*), the intellectual elite that rules society; (2) managers (*chefs d'entreprise*), who watch over society's daily operations; (3) subordinates (*bon sous-ordre*), who help the managers maintain society; and (4) workers (*les plus courant travail*), who actually carry out the daily operations of society. To emphasize the importance placed on knowledge, the leader of the planet is given the title of professor (Richard-Bessiere 126).

The hierarchical and global aspects of Mars's society call to mind the ideas of the influential Comte de Saint-Simon (1760–1825). As Robert Gilpin points out when writing about the reforms of scientific policies of the Fourth Republic, Saint-Simon provided the inspiration for government planning using scientifically and technologically advanced ideas as well as the centralization of policy-making (177).

However, despite the advantages of advanced knowledge possessed by Mars, there still remains much room for improvement. The ideas of Saint-Simon also provide the answer to Mars's imperfection. What is missing in Mars's society from Saint-Simon's social organization are the king, a noble class, and a stratum devoted to the arts and letters. The exclusion of the first two is viewed as an improvement, while the last omission Richard-Bessière covers in a later novel and on a different planet, Venus. The advanced society on Venus is described as "truly marvelous and comparable to the Eden of our scriptures" (179). Richard-Bessière, following in the best tradition of Saint-Simon, views all branches of human knowledge as necessary for human progress. Emphasis on science and technology is good, but overemphasis on these two fields of endeavor at the cost of the arts and letters leads to a sterile society where human happiness is sacrificed for mere efficiency.

As mentioned, Richard-Bessière's treatment of Mars can serve as a paradigm for the theme of what is the appropriate scale or proportion for the application of science and technology in modernization policies. More science and technology is good, but there exist limits — the balancing of these two fields with the development of all other endeavors of human culture. In fact, Richard-Bessière would later introduce the ideas of Henri Bergson, incorporating other sources into his ideas of progress, and Jean Monnet, stressing the role of cooperation in modernizing endeavors, to demonstrate both the potential and limits (or guidelines) of science and technology.

The remaining writers to be covered follow the same Voltairian implementation of scale and proportion when discussing how much is too much (or too little) for a society to progress in the modern age.

B.R. Bruss takes sides in the cold war when he uses Mars as a warning about the dangers of using science and technology in absolute fashion. His three-novel series (but *S.O.S. Soucopes* most importantly) details the struggle for supremacy between Earth and Mars.

In the near future the United States and the Soviet Union are still the reigning superpowers and engaged in their bitter rivalry. The Martians, meanwhile, are planning to conquer Earth. The would-be conquerors from the Red Planet ally themselves with the USSR due to the latter's resemblance to the former's. Martian society is super-efficient, scientific and technological. All facets of life revolve around the smooth functioning of the state. The Martians reproduce like plants, possess no private property, live in groups in identical cubical buildings, and basically do nothing but eat, sleep, and work. In return there exists no sickness and everyone has a useful place in society. The workers, who comprise the vast majority of the population, carry out the ultimate act of dedication to social efficiency by voluntarily limiting their lifespan, willing to end their lives at age fifty when their usefulness is viewed as starting to decline.

Some of the Soviet elites begin to doubt the alliance with Mars. One such skeptic expresses his doubts:

> The Martians are of a very prodigious scientific intelligence. To this regard they are ... extraordinarily interesting, but that is all. I have not found in them the least trace of what could resemble a human sentiment. They never laugh. They never cry. They never suffer.... They do not know anything of what colors the life of men. They have nothing, absolutely nothing, which resembles art....
> Compared to them, even termites are monsters of individuality [112–113].

The United States and the dissident elements of the Soviet Union eventually unite and defeat the Martian threat. Despite the superefficiency of the war-making capabilities of Mars, it is the cooperation between the once dreaded rivals that saves the day.

Bruss agrees with Richard-Bessière about the dangers of overemphasizing science and technology. The Saint-Simonian dictum of a balanced knowledge base seems to echo in his works, for it is a united (unity is also a well-known Saint-Simonian desire — in his case, that of Europe), but humanistic Earth that uses science and technology (as well as those stolen from the Martians) to defeat their superefficient but unbalanced neighbors.

Kemmel, the first of the three writers who treats Mars as the savior or guardian of humanity, still dwells on the theme of balanced development of human knowledge. Furthermore, he focuses on just one aspect of new science and technology, nuclear weapons.

In *Je reviens de...*, a group of Earth people are taken into outer space by some mysterious aliens who eventually reveal themselves as Martians. The aliens first take the group to Mars to see what a true advanced civilization looks like and then to Venus, to show what happens when science is abused. The second planet from the Sun shared an identical history with the Earth's, but with one horrifying difference: its inhabitants destroyed themselves in an atomic war.

Both groups decide that the Earth people should return home so they can warn their fellow humans about the possible dangers from the misuse of atomic energy. A member of the Earth party asks the Martian if he could take back to Earth one of the Martians' scientific discoveries. One of the Martians says no. As he explains, "If intellectual and moral civilization does not penetrate ... as quickly as technical civilization, the latter is fatally employed ... to the destruction of the former" (Kemmel 180). Once more a balance in human development is stressed.

The group of Earth people is French, and one of them attempts (but fails) to warn the president of France about the Martians message. Neither the Soviet Union nor the United States is involved. This can be interpreted as Kemmel having France in mind when he warns against the development of atomic weapons and the isolation of scientific and technical culture.

It should be pointed out that during the time of the publication of Kemmel's novel, and while the Fourth Republic's Parliament vacillated over France's atomic policies, a small group of people in the Commissariat à l'Énergie Atomique (CEA), the military, and the government was actually planning the organization of a project for the development of an atomic bomb. By the end of 1956, before the release of Kemmel's novel, the question of whether or not France should have atomic weapons was for all practical purposes already decided. The year of Kemmel's novel's appearance, 1957, witnessed some public discussion on this matter and, on April 11, 1958, Prime Minister Gaillard signed an order calling for the detonation of an atomic bomb by early 1960 (Scheineman 182–91).

Had his novel come out a year or two later, it would be safe to say that the warning of the Martian cited above about the penetration of technical civilization would take on more urgency, this time arguing against research policy being in the hands of a small, secret group of government elites. Social progress should not be dictated from above. It must be shared by all.

Maurice Limat, the second writer of this group depicting Mars as Earth's savior, finds a different source of wisdom to save humanity from its abuses of science and technology. In *J'écoute l'univers*, a father and son discover that they have the power to create by thought alone. Eventually the son uses his newfound psychic power to save the Earth from an alien invasion.

It turns out that one of the few survivors of the lost race of Mars was responsible for giving these powers to the father and son. He reveals the existence of powers beyond the visible and how humans, having acquired some scientific knowledge, are unwise to think that they have discovered all that there is to be discovered. Furthermore, a person needs to believe that his or her powers will work. In the novel, the father is cautious and skeptical, while the son — being young and naïve — is not. So it is only the youth who can exercise this power. This provides a clue to how Limat views the role of new scientific and technological knowledge.

Limat does not go into detail about the Martian society, so one needs to look at his work as a whole to obtain a complete picture of his ideas on progress. His later novel *Moi, un robot*, expresses a complete outline of his solutions for dealing with science and technology in particular and for solving humanity's problems in general. (This title is not be confused with Isaac Asimov's *I, Robot*. Though Asimov's book came out in 1950, its French translation came out after Limat's title. So unless Limat was able to read the American edition, it remains doubtful that he borrowed the title of his novel from Asimov's work.)

For Limat, religion is the key. What is needed lies beyond human knowledge. The approach to religion he advocates is not anti-intellectual, but one arguing for the complete human being, which includes the body, mind, heart, and all three guided by the soul, which is created by God. This is nothing new. Limat's ideas mirror much of the tenets of the Christian Democracy movement in postwar France. These politically focused ideas can be found in the publications of the political party that attempted to work out these ideas, the Movement Républicain Populaire (MRP) (Barron, Capelle, Einaudi and Goguel, Fogarty, Irving).

Jimmy Guieu, the third and last of this group, and probably the most idiosyncratic of all of the Anticipation writers, provides the dissenting voice in this discussion of Mars as a reflection of human concerns over science and technology. Not only does he see Mars as the source and salvation of human progress, but he is also the only one who wants more science and technology and seemingly in an uncritical fashion.

Noted for his incorporation of his belief in unidentified flying objects (UFOs) as real and portraying flying saucers as their representative form, Guieu's stories argue that present human institutions impede scientific and technological progress.

In *Nous les Martiens*, Guieu incorporates the ideas of Charles Fort, who viewed the Earth and humanity as products of superior races from other worlds. Guieu's version of Fort's theme has Atlantis, with all of its scientific wonders, as having existed and been founded by people from Mars. The Martians are ancestors of the White and Red (Native American) races, with the former being the masters since they were the majority. From Venus came the originators of the Black and Yellow races, with the latter depicted as tyrants. These two races fought each other in a war which reduced Earth to barbarism, losing all scientific and technological knowledge and forcing humanity to learn everything all over again. (One should note here that this seemingly callous portrayal of the major human ethnicities can be found in several of Guieu's works [Lofficier and Lofficier 422].)

Most of Guieu's other stories go along this line. Humanity is presently so inept with its science that it would probably take outside sources, or even an attack from them, to encourage people to focus on science and technology more.

Can humanity achieve scientific greatness on its own? To answer that question, Guieu, in *L'Ere des Biocybs*, borrows from the ideas of J.D. Bernal and depicts future humans as devoid of all familiar characteristics of human nature. They have become pure intellectual beings supported by advanced technology. Guieu may have been very positive on science and technology, but his focus on human inadequacy in the face of new knowledge and power also produces in his novels a counterbalance of pessimism. As a result, he urges humanity to employ drastic and extreme measures with science and technology for its salvation.

Up to now the various views on the scale of science and technology to be applied to human progress have been presented. The potential of these two fields of endeavor are beyond anything the French have encountered before. So the question now becomes: Should there exist any limits on them? Four writers say yes and one says no. Even among the affirmative side there is no agreement on how to limit science and technology. In the end, it is the limited abilities of humanity, whether by nature or culture, that forces this issue to the forefront.

But now is time for the last word on this theme. Or better yet, now is the time for the last laugh.

What most readers remember from "Micromégas" is the story's last scene where the two aliens are laughing hysterically over a conclusion made by the human philosopher who claimed that, from reading Thomas Aquinas, the entire universe was made solely for humanity's benefit.

The work of the last Anticipation writer to be examined here, Gilles d'Argyre, can serve as the last word, or the last laugh, depending on how one wants to interpret his work.

Gilles d'Argyre is the pseudonym of the noted writer, anthologist, critic, and editor Gerard Klein. Supposedly he chose this particular name to indicate the type of writing he was doing, the surname being a corrupted form of the Latin and French words for money. Thus Klein becomes a writer who writes only for the money, producing whatever the publishing market needs and requires.

He published one novel by the end of 1960, *Chirurgien d'une planète*. The story tells of Earth's attempt to transform the environment of Mars into one in which humans can live, a process referred to today as terraforming. However, a terraformed Mars would enable the settlers on it to become more independent, thus threatening those powers (on both planets) who want Mars to remain dependent on Earth. The invention of a matter-transfer device makes a quick completion of the project almost inevitable.

The established powers prepare to strike back, only to be thwarted at novel's end when their actions are revealed to the world.

In this novel Mars is not the focus of the story, but remains in the background. The competing forces struggling to control its fate occupy the center of the novel.

The terraforming is overseen by a government project called l'Administration. The organization calls to mind how scientific organizations are set up France. Since Napoleonic times, whenever a new discipline emerges or a new project arises, an independent institution is established and is answerable only to the government in Paris. As time goes by, it turns into an organization that jealously guards its independent status, which includes protection from the government itself. So each scientific institution becomes an intellectual fiefdom unto itself. This situation is part of a larger tension that has existed in France since 1789, that of opposing tendencies between centralization and factionalism. The Fourth and early Fifth Republics initiated reforms of this arrangement (Gilpin 78–85).

If his novel is to be the last word on this theme of Mars as a measure of science and technology and progress, then the word is a positive one. Gerard Klein has said that when he wrote the novel, his main goal was to write a "Fleuve Noir" and nothing else. His adoption of his pen name should signify that. So the last word would be that entrenched economic and political interests stand in the way of progress, that science and technology will expand and diversify humanity. (I have written elsewhere that this novel is not a typical "Fleuve Noir" as Klein claims it is [Lyau 292–294]; not just in terms of theme and characters, but also in the use of certain terms and in its ending, which could be considered ambiguous. However, Klein reiterated to me when I met him in 1991 that he was trying to write a "Fleuve Noir" and nothing else.)

One can also make an argument that Klein has actually given, instead, the last laugh on this matter. Several hints can be said to emerge throughout the novel. First, his adoption of the pen name — beyond the notion that he is writing only for the money — is also alluded to in the novel. One of the antagonists in the story is named Jon d'Argyre, a super-wealthy person who will do anything to gain economic power. Is this mere coincidence?

Second and more substantive, is the use of the term l'Administration as the name for the great project that is terraforming Mars. As Demetre Ioakimidis, in a review of this novel, once noted: "Let us not omit to raise up again a delightful touch of humor: the organization charged with the execution of this grandiose operation of planetary surgery is designated with a name that is generally synonymous with disorder, red tape, loss of time, and of sticks in the wheels ... the Administration" (133–134).

Third and most important, is the role of the character Georges Beyle, who is assigned by science police to help with the project. All through the novel he displays an undiminished faith in science, technology, and human abilities. What makes his role in the novel noteworthy is that in the first half of the novel the main protagonist is actually a scientist named Archim Noroit, who developed the terraforming capabilities. As the novel progresses, the scientist becomes less prominent and Beyle takes over as chief protagonist. Beyle turns out to be the more enthusiastic of the two, so much so that he becomes fanatical. When he compares himself to Noroit, he points out, "Archim sees the practical side of the project. For me, it is another thing. It is a dream. A mad dream. It is pride (*orgueil*) that possesses us" (118). *Orgueil* in this case could easily be translated as arrogance or vaingloriousness.

Towards the end of the story Beyle is seriously injured from an attack on the project, but he retains his singular devotion. So intense is his devotion that a colleague of his observes, "This man is mad ... genial perhaps, but mad" (183). Beyle is confined to a wheelchair due to debilitating injuries suffered in an attack on the project. With his fanatic devotion he could easily be a foreshadowing of a more memorable wheelchair-bound fictional character, Doctor Strangelove, from the movie of the same name released four years later.

Beyle even expresses the last spoken words in the novel with an exuberant soliloquy in which he orates glowingly about humanity's destiny to explore, transform, and conquer the universe. Mars for him is only the first stepping-stone of humanity's outward reach. However, Klein closes the novel itself with the following lines: "Georges Beyle slept. And the dreams or nightmares which he makes, he cannot share with anyone" (188).

And so the novel ends on a somewhat ambiguous note. One can claim that having Beyle close the story is Klein's way of saying that this novel is not merely a story advocating human progress through science and technology, but rather a take-off on such a story. Either Klein is critically concerned about such an approach to progress or he is actually making fun of this type of story. The reader is left to decide which way to interpret this ending, and — as a consequence — the novel as a whole.

Meanwhile, it should be recalled, "Micromégas" also finishes at an open end. After the two aliens have finished their convulsive laugh at human provincialism and limitations, the one from Saturn gives to humanity a rare philosophy book which contains the explanation of the ultimate essence of the universe. However, when the book was finally opened, its pages were discovered to be blank. Here, too, the reader is left without a definitive answer.

Both Voltaire's short story and the Anticipation novels' views of Mars are explorations of scales and limits. Both are products of a combination of two literary traditions. Both are concerned about how humans would react to the challenges of newly discovered knowledge. And finally, both belong to a distinctly French tradition of examining ideas.

So much of the history of science fiction is seen through the lens of British and American developments. But there is a French approach to this branch of literature, often referred to by its supporters as the literature of ideas. This is not to state that French science fiction is a separate development, but it is a distinct one. Perhaps tracing the development of science fiction from Voltaire's short story, as opposed to Shelley's *Frankenstein* or Edgar Allan Poe's *The Narrative of Arthur Gordon Pym*, would lead to a deeper appreciation of the diversity that really exists in science fiction's history.

Works Cited

Argyre, Gilles d'. *Chirurgiens d'une planète*. Collection Anticipation, no. 165. Paris: Fleuve Noir, 1960.
Barron, Richard. *Parties and Politics in Modern France*. Washington, DC: Public Affairs Press, 1959.
Bruss, B.R. *S.O.S. soucoupes*. Collection Anticipation, no. 33. Paris: Fleuve Noir, 1954.
Capelle, Russell B. *The MRP and French Foreign Policy*. New York: Frederick A. Praeger, 1963.
Einaudi, Mario, and François Goguel. *Christian Democracy in Italy and France*. Notre Dame: University of Notre Dame Press, 1952.
Fogarty, Michael P. *Christian Democracy in Western Europe 1820–1953*. Notre Dame: University of Notre Dame Press, 1957.
Gilpin, Robert. *France in the Age of the Scientific State*. Princeton, NJ: Princeton University Press, 1968.
Guieu, Jimmy. *L'ere des Biocybs*. Collection Anticipation, no. 160. Paris: Fleuve Noir, 1960.
_____. *Nous les Martiens*. Collection Anticipation, no. 31. Paris: Fleuve Noir, 1954.
Ioakimidis, Demetre. "Ici, on desintegre!" *Fiction*, December 1960.
Irving, R.E.M. *Christian Democracy in France*. London: George Allen and Unwin, 1973.
Kemmel. *Je reviens de ...* Collection Anticipation, no. 84. Paris: Fleuve Noir, 1957.
Limat, Maurice. *J'écoute l'univers*. Collection Anticipation, no. 154. Paris: Fleuve Noir, 1960.
_____. *Moi, un robot*. Collection Anticipation, no. 170. Paris: Fleuve Noir, 1960.
Lofficier, Jean-Marc, and Randy Lofficier. *French Science Fiction, Fantasy, Horror and Pulp Fiction: A Guide to Cinema, Television, Radio, Animation, Comic Books and Literature*. Jefferson, NC: McFarland, 2000.
Lyau, Bradford. "Technocratic Anxiety in France: The Fleuve Noir 'Anticipation' Novels, 1951–60." *Science-Fiction Studies*, November 1989.
Richard-Bessiere, F. *Les conquérants de l'univers*. Collection Anticipation, no. 1. Paris: Fleuve Noir, 1951.
_____. *Retour du Météore*. Collection Anticipation, no. 2. Paris: Fleuve Noir, 1951.
Scheineman, Lawrence. *Atomic Energy in France under the Fourth Republic*. Princeton, NJ: Princeton University Press, 1965.
Wade, Ira O. *Voltaire's* Micromégas: *A Study in the Science, Myth, Art*. Princeton, NJ: Princeton University Press, 1950.

IS MARS HEAVEN?
The Martian Chronicles, Fahrenheit 451, and Ray Bradbury's Landscape of Longing
Eric S. Rabkin

In the fall 1948 issue of *Planet Stories*, Ray Bradbury published a short story that the editor didn't even mention on the cover, yet this story, with the exclamatory title "Mars Is Heaven!," went on to become one of the most famous science fiction stories of all time. The first Earth expedition, comprising white men from Ohio, lands on Mars in 1960 on a "lawn of green grass" before a Victorian house in the midst of a small Midwestern town. Their instruments tell them the atmosphere is just like Earth's. The men want to run out to explore what one calls "Good old Mars!" (322). Captain John Black cautions his men against this impossible reincarnation of their youth, but nostalgia cannot be long resisted. They run off to what Black recognizes as Green Bluff, Illinois, his own childhood town. (Bradbury's fictional setting, which perhaps reflects his own native Waukegan, Illinois, exists in no real atlas but turns out to be a green bluff — that is, a verdant deception — indeed.) Even before they meet people and hear voices, through a window the crew see an untended piano on which sits the sheet music for "Beautiful Ohio." Then the people appear, all departed loved ones. The crew reel with joy, dismiss their own puzzlements, and scatter to the homes of those they had once mourned. Even Captain Black retires to his family, incredulous that Mars could possibly realize all their hopes, but seduced nonetheless. As he finally settles down to sleep in a bed shared with his now undead brother, Black thinks, just perhaps, this is all a mind trick, a ploy by telepathic Martians to disarm invading Earthmen. He looks up just in time to see the Martian impostor stab him to death. The narrator informs us that elsewhere the whole crew is slaughtered.

Is Mars heaven? Bradbury's clear reply in 1948: No. Heaven, or the vision of heaven shaped from our own longing, is a fatal snare. And Mars is a landscape of longing.

According to William G. Contento's *Index to Science Fiction Anthologies and Col-*

lections, Bradbury is the single most reprinted science fiction short story writer in the world. Bradbury's most reprinted story is "A Sound of Thunder" (1952), the fictional source for the name of the real-world phenomenon called "The Butterfly Effect." In that story, a time-traveling tourist barely steps off the path in the Jurassic and irrevocably changes all the millions of years that follow. As in "Mars Is Heaven!" we see here Bradbury's recognition that while the past may lure us, stepping onto the lawns of the past may lead to our destruction in the present. His next most reprinted story is "There Will Come Soft Rains" (1950), an ironic elegy in which household mechanicals, both built-in systems and mobile robot appliances, mindless of the sudden death of humanity by atomic war, try to perform their programmed functions until, bumping into each other and misperceiving their environment, they, the last vestige of our ingenuity, fail miserably, bursting into flame, finally consumed like the rest of Earth civilization by fire (*Martian Chronicles* 166–72). No matter how loyal and well designed our technology may be, Bradbury seems to say here, our faith that our technology reliably maintains the world we want cannot be sustained. Bradbury's third most reprinted story is "Mars Is Heaven!" or, more accurately, that story and a version of it that appears in *The Martian Chronicles* (1950).

The Martian Chronicles, published May 7, 1950, is a milestone in the history of science fiction in specific and American literature in general. Today's readers may not appreciate the widespread ignorance and disparagement of science fiction then. About eight months before the book's publication, on August 28, 1949, pioneering SF writer and editor Donald Wollheim published a full-page essay in the *New York Times* called "The Science-Fiction Novel." Most of it is devoted to basic definition, starting with distinguishing science fiction from "say, a fictionalized biography of Marconi or Pasteur." About ten months after Bradbury's book's publication, on February 17, 1951, James B. Conant, erstwhile chemist and revered president of Harvard University, complained in that same newspaper, "It is a great reflection on us teachers of science at the university and the high school level that we have apparently been unable to put across to students, who have not majored in the sciences, just what science is about.... The public acceptance of science fiction or fiction disguised as science is our indictment." The *Times* did print capsule reviews of newly published science fiction novels from time to time, but began doing so only on September 18, 1949, when movie critic A.H. Weiler first bothered to give a brief review of Jack Williamson's 1948 novel *The Humanoids*, along with reviews of another novel (*The Big Eye*, by Jack Ehrlich) and an anthology (*The Best Science Fiction Stories—1949*, edited by E.F. Bleiler and T.E. Dikty). But *The Martian Chronicles* was something special.

On August 5, 1951, Harvey Breit published "Talk with Mr. Bradbury." As far as I have been able to discover, this is the very first interview that the *Times* ever ran with an acknowledged SF writer who interested the newspaper precisely because he was an SF writer. It begins by noting, "There is this genre, getting quite a play these days, known as science fiction." After a bit of joking about how Bradbury did not arrive in New York "in a space ship," Breit asks his first question. He could not have known then that Bradbury's resistance to some technologies would become so famous that *Time* magazine would spread the news that at the age of sixty-two (1982) Bradbury quite nervously took his very first airplane ride ([untitled notice]). "First off, before defining the field," Breit wondered back in 1951, "how did Mr. Bradbury come to be writing science fiction? 'I'm not only a science fiction writer,' Mr. Bradbury replied." Bradbury claims to work in three fields: "smalltown [sic] life, ... science fiction and fantasy." He distinguishes SF as being somehow possible, and uses his own story "The Pedestrian" as an example. In that story's future America, in which machines have made life relentlessly easy, a strolling man wanting just to breathe unconditioned air and see the world without an intervening television screen is accosted by robot traffic cops who are so incapable of understanding the man's desires that, as Bradbury says, "the pedestrian gets taken to an insane asylum." Fantasy, on the other hand, Bradbury calls "the improbable," such as "a dinosaur appearing in the streets of New York." Bradbury doesn't define small-town life as a genre, but when asked about the influences on his own writing, the first artist he cites is Sherwood Anderson.

Like Anderson's *Winesburg, Ohio* (1919), *The Martian Chronicles* is a composite novel (Rabkin, "Composite"). The chronological ordering of Bradbury's chapters, the creation of eight new chapters woven among the seventeen previously published pieces, and some subtle yet significant revision, including that for what had been "Mars Is Heaven!" demonstrate that the book is no mere collection. Taken as a whole, it traces symbolically powerful episodes in Earthlings' migration to Mars. The book is profoundly American. "The Green Morning" chapter, for example, is a version of the story of Johnny Appleseed. Here, though, the counterpart of that legendary hero is not merely planting trees but, by their presence, enriching the air with oxygen. That the Appleseed character could walk the Martian landscape without breathing apparatus even before planting the trees is just as improbable as a living dinosaur in New York. That the trees grow overnight like "gigantic beanstalks" (76) doubly suggests that the story, clearly not SF, is not only fantasy but in particular the variety of fantasy we call fairy tale (Rabkin, "To Fairyland"). The chapter called "Way in the Middle of the Air" tells how American blacks escape from racism by taking a rocket to Mars, leaving their excess belongings behind like a Civil Rights version of the "Gone to Texas" sign people once tacked to their doors when they would "light out for the Territory," to quote from the last paragraph of the memoirs of Nigger Jim's companion, Huckleberry Finn (Twain 539). In the "Usher II" chapter, an "Investigator of Moral Climates" (107), the bluenose agency that bans imaginative authors like Poe, is welcomed to a fantastic "masque" (a dress-up ball as in Poe's "The Masque of the Red Death") in which the guests on Mars and finally the investigator himself succumb to literary devices that fans of Poe have all seen before: premature burial, immurement in the cellar, being stuffed up a chimney, and finally the whole "house" and scene collapsing into a "tarn" like that in the last paragraph of Poe's precursor story, "The Fall of the House of Usher" (Poe 268).

Near the beginning of *The Martian Chronicles*, when the first Earthman, Nathaniel York, comes to Mars, he is sensed far off. A Martian wife named Ylla begins unaccountably to sing "Drink to me only with thine eyes" (5). Her husband, disturbed to see her moved, takes a weapon that fires "golden bees," meets the descending ship in the next valley, and kills York. When the second expedition arrives with four men, a Martian crowd traps them by telepathy in an insane asylum. The third expedition, called in *The Martian Chronicles* simply "The Third Expedition," is Bradbury's revision of "Mars Is Heaven!" In each of these chronicle entries, science is left behind; the imagination wins. Then, between the third and fourth expeditions, a childhood disease, supposedly carried unwittingly by the Earthmen, all but destroys the Martians. Thus the rest of the novel considers the attempts at an American frontier expansion without the American original sin of land theft and genocide. There are a few chapters that remind us of the West's (that is, Mars's) relation to the East (that is, Earth), as when retirees move to Mars. We have critiques of capitalism, as in the chapter called "The Off Season." But the novel's main complaint is aimed at technology itself.

Ultimately Earth consumes itself in atomic war. Those few pioneers who do not return to fight for their countries now wander a depopulated Mars. The penultimate chapter is "There Will Come Soft Rains," a critique not only of atomic weaponry but, like "The Pedestrian," which is not in this novel, a critique of turning our lives over to robots, to technology. Something different happens in "The Million-Year Picnic," the very last chapter of this composite novel. An Earth family comes to Mars in a private rocket. The father had planned this escape from the expected atomic conflagration, and as he leads his wife and sons inland on a canal boat, by remote control he destroys their rocket so that "evil men" (179), should they come to Mars again, will not be able to find them. They await another refugee family, this one with daughters. Around a campfire, the sons clamor to see the Martians their father had promised to show them. He takes them back down to the canal and asks them to look into the water. "The Martians stared back up at them for a long, long silent time from the rippling water..." (181). Thus the novel ends, as Crèvecoeur had said European emigration to America ended, by the glorious transformation of the sinful individual into "this new man ... who leaving behind him all his ancient prejudices and manners receives new ones from the new mode of life he has embraced" (34). Then, an American; now, a Martian, which is to say an American who has a second chance untainted with imperialism and having escaped the 1950s' technology-branded fear of nuclear self-immolation.

It is little wonder that a book like this was the first to gain a putatively SF author extended treatment in the *New York Times*. In 1954, when the National Institute of Arts and Letters honored Bradbury for his contributions to American literature, the work cited was *The Martian Chronicles*. It won, that same year, the second annual gold medal of the prestigious Commonwealth Club of California. While one might argue that Bradbury was only occasionally a science fiction writer, undoubtedly Conant had Bradbury's seductive fantasy, and those works it represented, in mind when he called science fiction his own professional indictment. For an American audience in the early 1950s, *The Martian Chronicles* offered literary continuity and mythic redemption without any real science at all.

The Martian Chronicles also confirmed the mythic grounding of American culture in small-town life. In the opening chapter, "Rocket Summer," an "Ohio winter" in a "small town" magically turns to summer through the heat of a rocket rising from its nearby launching pad. "The rocket made climates" (1) — that is, charged the very air shared by everyone. From that chapter on, we see small towns as the models of community in the chapters called "The Third Expedition" (which is the revision of "Mars Is Heaven!"), "The Settlers," "Night Meeting," "The Martian," "The Silent Towns," and "The Long Years," among others. It is no wonder then that this suddenly lionized writer of fantasy but called a writer of science fiction, when he served as a consultant to Walt Disney in the design of Disneyland, approved the main entrance to "The Happiest Place on Earth" going through an old railway station to the foot of "Main Street U.S.A.," the portal to "Frontierland," "Fantasyland," and "Tomorrowland." Disney's fantasy small town with its main street anchored by the train station nearly reproduces the map opposite the title page in *Winesburg, Ohio*, the first great American composite novel and one that Bradburg admired.

Young Bradbury was christened a Baptist, but was "a self-confessed agnostic in his teens" (Dimeo 157). Still, in his middle years, he acknowledged a continuing moral concern. What did that mean to Bradbury? "Now," he wrote, "very late in the scroll of Earth, phoenix man, who lives by burning, a true furnace of energy, stoking himself with chemistries, must stand as God. Not *represent* Him, not *pretend to be Him, not deny Him, but simply, nobly*, and frighteningly *be* Him" (quoted in Dimeo 159). Steven Dimeo calls Bradbury "clearly pantheistic" (159). Bradbury lists himself in *Contemporary Authors Online* as Unitarian Universalist. One may have moral concerns, of course, without recourse to any god much less a Christian god, but given Bradbury's upbringing and interests, it is surprising that the *Modern Language Association Bibliography*, when queried for Bradbury and religion, returns only one hit.

Works like Walter M. Miller's *A Canticle for Leibowitz* and Arthur C. Clarke's "The Nine Billion Names of God" demonstrate that science fiction can explicitly and powerfully explore the relations between science and religion. Bradbury does not do that. On his own website, Bradbury notes that *Farewell Summer* (published in 2006 but begun in 1944) "was a response to my ganglion and my antenna. I do not use my intellect to write my stories and books; I have a gut reaction to the things that my subconscious gives me. These are gifts that arrive early mornings and I get out of bed and hurry to the typewriter to get them down before they vanish" ("In His Words"). Statements of faith can be written this way, but reasoned theology cannot.

Bradbury ended his formal education with high school. Of course, this hardly means that his education ended. Quite the contrary. His avid reading, and particularly his affinity for American letters, shows everywhere in his writing. Naming the Martian housewife "Ylla," for example, may be one of those subconscious gut reactions Bradbury says drives his writing. "Yillah," with its slightly different spelling, is the unreachable white maiden, a would-be religious sacrifice, the quest for whom motivates the episodic action in Herman Melville's novel *Mardi* (1849). I don't know that Bradbury ever read this now little-read precursor to *Moby-Dick* (1851), but the character fits, the spelling is off enough to suggest little intentional scholarship, and we know Bradbury has an artistic affinity for Melville since Bradbury wrote the screenplay for the award-winning film of Melville's masterpiece.

On the other hand, Bradbury, whose lyric control of language always astounds, may have been playing a conscious word game by alluding subtly to *Mardi*. *Mardi* means "Tuesday" in French. Mardi is the weekday named for the Roman war god, Mars. In "The Earth Men," the chapter of *The Martian Chronicles* recounting the second expedition, when Captain Williams tells the first native he meets that he has come from Earth to Mars, she snappishly — and telepathically — replies, "'This is the planet Tyrr ... if you want to use the proper name'" (17). Tyr is an Old Norse god of war (Davidson 238). The Old English version of that god is Tiw, from which we English speakers get "Tuesday," which the French call *mardi*.

Whether or not consciously, Bradbury's imagery, like his allusions, clearly draws from his passionate reading yet stamps his work as his own. In all of Western culture, from the book of Genesis to the present, imagery of light and sight has referred to knowledge and intellect. *I'm sure you see what I mean; this is a useful perspective; my vision is not cloudy but clear; put aside those rose-colored glasses if you seek true insight. Do I have to draw you a picture?* Too much knowledge is overwhelming, be it the light that blinds Saul of Tarsus and leads to his transformation into Saint Paul or the terrible anagnoresis of Oedipus that leads him to blind himself. Plato's parable of the cave equates light with knowledge. And we must remember that when the former cave-dwellers return with new insight to share with their stationary fellows, they are thought to be insane and are killed. Light and sight are knowledge, and too much can cost us our identities, our minds, even our lives. At the end of *Frankenstein*, the monster vows to go make himself a "funeral pile," the tragic bookend to the lightning strike that first animated him, so that "soon these burning miseries will be extinct" (223).

If light imagery reflects intellect and Bradbury reports that his writing comes from the gut, it fits that the imagery in *The Martian Chronicles*, although it certainly includes light, much more strikingly features music. Intellect works on the knowledge that comes through the eyes, precise knowledge that can be seen or shut out. Music is atmospheric, climatic if you will. Yes, the rocket in "Rocket Summer" launches, but rather than show the light, the narrator shows us people shucking off their heavy clothing. What we hear as music, at least its point of origin, is much less precise than what we see. Bradbury's Ylla has no idea where the song in her mind originates. In "The Summer Night," the chapter just before four humans arrive on Mars for the second expedition, the lives of the Martians' "little towns" (14) are disturbed by what we recognize as Earth songs that erupt unbidden from the throats of children and adults alike. In "The Long Years," someone believing himself the last man on Mars is drawn awkwardly toward the character who may be the last woman by the strains of "Genevieve, Sweet Genevieve." Even when music is unmentioned, the language is lyric (like the sound of a lyre), and details like the sheet music for "Beautiful, Ohio" remind us of music. Music makes climates, but it does not make logical arguments.

The second most famous book by this nominal science fiction writer is *Fahrenheit 451* (1953). A "fireman" named Montag, whose job in this dystopian future is to burn all books, begins to wonder if books, which some people willingly die for, actually may be worth preserving or even reading. His wife, a drone, lives for those television screens, here a sham interactivity and part of the thoroughly mechanical environment that Leonard Mead, Bradbury's "Pedestrian," shunned. Clarisse, a vibrant girl who admits to reading, captivates Montag, as eventually does Faber, forty years retired from his post as the last English professor.

This is how the book begins:

> It was a pleasure to burn.
> It was a special pleasure to see things eaten, to see things blackened and *changed*. With the brass nozzle in his fists, with this great python spitting its venomous kerosene upon the world, the blood pounded in his head, and his hands were the hands of some amazing conductor playing all the symphonies of blazing and burning to bring down the tatters and charcoal ruins of history. With his symbolic helmet numbered 451 on his stolid head, and his eyes all orange flame with the thought of what came next, he flicked the igniter and the house jumped up in a gorging fire that burned the evening sky red and yellow and black. He strode in a swarm of fireflies. He wanted above all, like the old joke, to shove a marshmallow on a stick in the furnace, while the flapping pigeon-winged books died on the porch and lawn of the house. While the books went up in sparkling whirls and blew away on a wind turned dark with burning.
> Montag grinned the fierce grin of all men singed and driven back by flame.
> He knew that when he returned to the firehouse, he might wink at himself, a minstrel man, burnt-corked, in the mirror. Later, going to sleep, he would feel the fiery smile still gripped by his face muscles, in the dark. It never went away, that smile, it never ever went away, as long as he remembered [3–4].

One reads of burning here, of a fire extinguishing the light that could have shone from books, but that light is absent and the fire here is less light than the heat of passion, with the nozzle in Montag's frenzied fists and his hose a great python. Pythons, by the way, are not venomous, any more than all varieties of paper ignite at 451 degrees Fahrenheit. The gut guides Bradbury's writing brilliantly — that is, his writing shines — but the shining writing, as he said himself, at bottom does not reflect intellect, no matter how learned his prose may sometimes be. Even in this opening description of burning, Montag is a "conductor playing all the symphonies" and, like the cold distant atomic conflagration of Earth barely seen from Mars in *The Martian Chronicles*, such light as there exists made "a wind ... *dark* with burning." Montag imagines seeing himself not as a fireman but as a black-faced "minstrel man." While light and sight imagery are powerful in most science fiction, in Bradbury's two most renowned works, sound and music dominate, not intellect but the gut.

Bradbury's writing, whether consciously or not, connects with the wide world of writing, not only in the maps of Disneyland and Winesburg, or the composite structures of *Winesburg, Ohio* and *The Martian Chronicles*, but in systems of symbols and bits of plot. Nathaniel Hawthorne's short piece called "Earth's Holocaust" (1844) recounts a future — or a parable, the narrator will not commit — in which all "accumulation of worn-out trumpery" (179) is to be immolated. First come signs of rank, then office, but finally books, and all are heaped onto the fire. As the viewpoint character watches it all go, the last to burn is the Bible. Indeed, when the ashes are combed, we find that book singed, the marginal notes burned clear, but the glorious text itself still present. The word "Holocaust," we should remember, before its association with Nazi death camps, meant "burnt offering." It is cognate with "cauterize," to heal with fire. Hawthorne's viewpoint character is told not to fear, for the real gems live on in the human heart (196).

At the end of *Fahrenheit 451*, Montag, for having kept and been attracted to some books, must flee for his life. He decides to join an exile society Faber has told him of in which each person *is* a book, that is, each person has committed a book to memory. "We're nothing more than dust jackets for books," (153), Montag is told. Although someone says Montag "look[s] like hell" (154), someone else chides that you "don't judge a book by its cover" (155). Montag chooses to be parts of Ecclesiastes and of Revelation (165). The outdoor community of book people is led by Granger, a word that means farmer. An alert reader might recall the imagery of the novel's beginning, with the "great python" perhaps standing in for the serpent and the burning "flapping pigeon-winged books [that] died on the porch and lawn of the house" perhaps predicting the ascending dove of the Holy Ghost. "[T]he books went up in sparkling whirls," and at the end Montag *is* the book, just as Bradbury has said that "phoenix man" must be God. Did the sparks fly all the way to Mars?

Montag is German for Monday, the day before Tuesday, Mardi, Yllah's domain. Montag is also the brand of typewriter paper Bradbury used for writing *Fahrenheit 451*, and Faber & Faber the pencils with which he edited his manuscript. But whether these name choices were truly "subconscious," as Bradbury has written (Afterword 173) or not, it is clear that Faber, meaning "craftsman" in Latin, whether or not coincidentally, here honors the making of a book, and the environment in which books live is managed by Granger, the farmer just outside town.

Whether from gut or from intellect, Bradbury made crucial changes in transforming the short story "Mars Is Heaven!" into "The Third Expedition" in *The Martian Chronicles*. In addition to dropping the original title and moving the time forward forty years (a symbolic number; just ask Noah or the Hebrews in the desert), he changed the text. Consider the last three paragraphs. Those in italics are from the 1948 original, those without from the 1950 composite novel.

> *The coffins were lowered. Somebody murmured about "the unexpected and sudden deaths of seventeen fine men during the night—"*
> The coffins were lowered. Someone murmured about "the unexpected and sudden deaths of sixteen fine men during the night—"
>
> *Earth was shoveled in on the coffin tops.*
> Earth pounded down on the coffin lids.
>
> *After the funeral the brass band slammed and banged into town and the crowd stood around and waved and shouted as the rocket was torn to pieces and strewn about and blown up.*
> The brass band, playing "Columbia, the Gem of the Ocean," marched and slammed back into town, and everyone took the day off.

Bradbury has, of course, made the language more lyric. In addition, he has modified the psychology. In the free-standing original, the Martians are joyful in their destruction, joyful, we imagine, in a Martian way. In the novel, this early chapter occurs after two earlier expeditions. As the balance of Earthmen to Martians becomes more even, the Martians not only work their telepathy on the Earthmen but the Earthmen unknowingly taint the Martians, as we have seen in the Martians' unwitting enthrallment to human song. Where before there had been seventeen dead men, which included a man who died in transit and never walked Martian soil, now there are only sixteen, the ones the Martians themselves killed. Where before the Martians cut loose in their final destruction of material from Earth, now they continue to be shaped by Earth ideas of work schedules, bands, and music. No, Mars is not heaven at the end of the third expedition because it is becoming, even if only musically, shaped by Earth.

By the end of *The Martian Chronicles*, we find a family from Earth that has given themselves up to Mars. "The Million-Year Picnic" (1946), the short story used as the last chapter in the book, was published first. One may guess that, in Bradbury's own way, he wrote the other pieces aiming to culminate with this one. The father destroys the rocket not only so that "evil men" cannot find the family but because on Earth "Science ran too far ahead of us too quickly, and the people got lost in a mechanical wilderness ... that way of life proved itself wrong and strangled itself with its own hands.... [However, i]t would have been another century before Mars would have been really poisoned by the Earth civilization" (179–181). At that point the father directs his children to look at the Martians, themselves, reflected in the canal water.

In both *The Martian Chronicles* and *Fahrenheit 451*, war cleanses. Does this burn our sins and bring us a new Eden? I think not. Granger was the right name for the leader of the book people. The consequences of the Fall are childbirth, labor (particularly including agriculture), and death. Bradbury's Mars finally welcomes boys and girls so that there will be childbirth; he expects there to be labor, but labor only with appropriate, scaled-back technology; and as for death, well, that comes for the dust jackets, but not for the books. In *The Martian Chronicles*, Ray Bradbury will live forever and generations will say, reading him in whatever form, that there the American myth comes clean again, small town life welcomes us all, and his Mars, undoubtedly, is heaven.

Works Cited

Anderson, Sherwood. *Winesburg, Ohio*. 1919. Reprint, New York: Viking, 1969.
Bradbury, Ray. *Fahrenheit 451: The 50th Anniversary Edition*. New York: Ballantine, n.d. (orig. pub. 1953).
_____. "In His Words." Web. September 1, 2007. http://www.raybradbury.com/inhiswords02.html
_____. "Mars Is Heaven!" (orig. pub. 1948). *The Science Fiction Hall of Fame*. New York: Orb, 1990.
_____. *The Martian Chronicles*. New York: Bantam Books, 1950.
_____. "The Pedestrian" (orig. pub. 1951). *Bradbury Stories: 100 of His Most Celebrated Tales*. New York: HarperCollins, 2003.
_____. "A Sound of Thunder" (orig. pub. 1952). *Bradbury Classic Stories 1: Selections from* The Golden Apples of the Sun *and* R Is for Rocket. New York: Bantam, 1990.
Breit, Harvey. "Talk with Mr. Bradbury." *New York Times*, August 5, 1951.

Clarke, Arthur C. "The Nine Billion Names of God" (orig. pub. 1953). *The Nine Billion Names of God: The Best Short Stories of Arthur C. Clarke*. New York: New American Library, 1967.

Contento, William G. *Index to Science Fiction Anthologies and Collections*. http://www.philsp.com/homeville/isfac/0start.htm

Crèvecoeur, J. Hector St. John de. *Letters From an American Farmer*. Belfast: James Magee, 1783. Eighteenth Century Collections Online. Web. http://catalogue.nla.gov.au/Record/3203876

Davidson, H.R. Ellis. *Gods and Myths of Northern Europe*. Baltimore, MD: Penguin, 1964.

Dimeo, Steven. "Man and Apollo: Religion in Bradbury's Science Fantasies." *Ray Bradbury*. Eds. Martin Harry Greenberg and Joseph D. Olander. New York: Taplinger, 1980.

"Dr. Conant Decries Science Fiction Rise." *New York Times*, February 17, 1951.

Hawthorne, Nathaniel. "Earth's Holocaust" (orig. pub. 1844). Ed. Alfred Kazin. *Selected Short Stories of Nathaniel Hawthorne*. New York: Fawcett, 1966.

Melville, Herman. *Mardi and A Voyage Thither: An Allegorical Romance*. 1849. Reprint, New York: Capricorn, 1964.

_____. *Moby-Dick or, The Whale*. 1851. Reprint, New York: Bobbs-Merrill, 1964.

Miller, Walter M., Jr. *A Canticle for Leibowitz*. New York: Bantam, 1959.

Plato, *Republic*. Trans. Paul Shorey. *Plato: The Collected Dialogues*. Eds. Edith Hamilton and Huntington Cairns. Princeton, NJ: Princeton University Press, 1961.

Poe, Edgar Allan. "The Fall of the House of Usher" (orig. pub. 1839). *The Portable Poe*. Ed. Philip van Doren Stern. New York: Viking, 1945.

Rabkin, Eric S. "The Composite Novel in Science Fiction." *Foundation* 66 (1996).

_____. "To Fairyland by Rocket: Bradbury's *The Martian Chronicles*." *Ray Bradbury*. Eds. Martin Harry Greenberg and Joseph D. Olander. New York: Taplinger, 1980.

"Ray Bradbury." *Contemporary Authors Online*. May 2, 2008.

Shelley, Mary. *Frankenstein, or The Modern Prometheus*. Ed. M.K. Joseph. 1818. Reprint, New York: Oxford University Press, 1969.

Twain, Mark. *The Adventures of Huckleberry Finn*. *The Portable Mark Twain*. Ed. Bernard DeVoto. 1884. Reprint, New York: Viking, 1968.

[untitled notice]. *TIME*. 8 November 1982: 53.

Weiler, A.H. "Out of This World." *New York Times*, September 18, 1949.

Wollheim, Donald. "The Science-Fiction Novel," *New York Times*, August 24, 1949.

RE-PRESENTING MARS
Bradbury's Martian Stories in Media Adaptation
Phil Nichols

Introduction

Conflict with landscape is common in science fiction (Rose 36–37), and Ray Bradbury's *The Martian Chronicles* is a classic illustration of this. Although many of the stories in the book find their central conflict elsewhere, a developed theme throughout the *Chronicles* is humankind's struggle to fit into a landscape of embedded memories and echoes of a prior civilization. The conflict is resolved in the final chapter by one Earth family staring the canals of Mars in the face, recognizing themselves as transformed by the planet's influence into Martians. The book fulfills Lomax's characterization of the role of landscape in SF, which is "to dramatize the need for a new totalizing myth in a world fragmented by the loss of the old" (253).

At the same time, Bradbury's Mars is a nostalgic reflection of Earth, echoing Said's and Schama's observations about the intersection of memory and "human space" (Said 175; Schama 120): Depictions of fixed landscapes are unstable, since the act of depiction involves an act of invention which in turn depends on the culture and experience of the creator of the landscape.[1] When Bradbury's text is subsequently amended through adaptation to other media, whether by an adapter or by the author himself, the landscape, both natural and built, can become highly unstable, and human interaction with the landscape in particular can shift such that a new, distorted light can be shone on the source text.

Bradbury is one of the most widely adapted of all American writers. His texts have an allure to adapters which sometimes proves harmful to a coherent adaptation as adapters struggle to find an appropriate relationship with the text. A study of Bradbury's self-adaptations reveals a different struggle, as the author and the original text seemingly do battle to achieve a controlling influence over the new media text. This chapter shows how the struggle for control can alternately harm and refresh an adaptation through its effect on the constructed landscape, and suggests that self-adaptation is an important but overlooked special case of adaptation.

I shall focus on just one story from *The Martian Chronicles*, "And the Moon Be Still as Bright," one of the most frequently adapted Bradbury stories (for a summary of adaptations of this and other stories from *Chronicles*, see table 1). After an analysis of the original text, I discuss its appeal to various adapters and some of the difficulties that arise from this. Finally, I examine Bradbury's own illuminating attempts at adapting the story for stage and screen.

"And the Moon Be Still as Bright"

"And the Moon Be Still as Bright" was first published in 1948 in *Thrilling Wonder Stories*. As a stand-alone story, it dramatizes a conflict of idea and attitude between crewmembers on a mission to Mars. Introspective crewmember Spender sets himself apart from the rest of the crew, attempts to gain respect for the extinct native race, and tries to stop the colonization. Spender adopts the identity of a Martian, exacting revenge on the Earthmen for their defiling of the planet. Although reaching something of a meeting of minds with his captain, he is unable to thwart the colonization, and the story ends with Spender being killed by Captain Wilder and buried in a Martian tomb. Spender had wanted to be a Martian, and in death he achieves his wish. There is a strong sense that Wilder will inherit some of Spender's attitudes towards Mars, and become something of a guardian of the old ways.

Bradbury has been accused of creating nothing but stereotypes in this story (Grimsley 1239): Spender is the idealist, Wilder the pragmatist. I would argue, however, that through Spender's interior monologue we actually get an understanding of a character who is affected by what he experiences, and who is allowed to grow.

The story has elements of "landscape as conflict," and Spender enacts this conflict as a personal struggle. Where Spender believes "the Martians must hate us," Captain Wilder sees evidence all around of intelligence and culture. Other crewmembers see the planet as an open space for partying and trashing. The Martian landscape is a Rorschach test for each person who looks at it. But the story also has dramatic conflict between characters, and it is probably this that has led to its popularity in adaptation; there have been eight adaptations to date.

Bradbury (*Martian Chronicles* unabridged reading) reports that he wrote "And the Moon" under the influence of Joseph Wood Krutch's *The Modern Temper*, which he was reading around 1946–47.[2]

Krutch writes: "Time was when the scientist, the poet and the philosopher walked hand in hand. In the universe which the one perceived the other found himself comfortably at home. But the world of modern science is one in which the intellect alone can rejoice" (12).

Bradbury's Spender paraphrases Krutch when he casts the Martians as still in that earlier mode: "They blended religion and art and science because, at base, science is no more than an investigation of a miracle we can never explain, and art is an interpretation of that miracle. They never let science crush the aesthetic and the beautiful" (*Silver Locusts* 88).

Krutch is also paralleled in the antagonistic characters in the story, Gibbs and Parkhill, "absorbed in the processes of life for their own sake ... conquering without asking for what purpose they conquer." Krutch's outlook permeates "And the Moon" and underpins Spender's and Wilder's attempt to reconcile themselves with the Martian landscape.

When brought into *The Martian Chronicles*, first published in 1950, "And the Moon" becomes a pivotal episode. In the context of the book, it functions as a turning point: It is here that the Martians are annihilated, and here that an Earthman first entertains notions of becoming Martian, foreshadowing the conclusion of the novel. In finding a place for the story, Bradbury carries out his first adaptation of the material, as he makes small but significant adjustments to the text.

The most substantial insertion is a sequence in which crew member Hathaway returns from a mission with news that he has founds dozens of dead Martian towns, with signs of a very recent extinction of the natives. The extinction is due to a previous mission, which unknowingly infected the Martians with chicken pox. This insertion — a matter of a few paragraphs — has an overwhelming influence on the narrative of "And the Moon":

a. it provides a link to the previous chapter of the *Chronicles*, "The Third Expedition"
b. it turns the disaster at the end of "The Third Expedition" into a pyrrhic victory;
c. it provides Spender with an objective underpinning for his hostility to humankind, and a direct motive for revenge. Coupled with small excisions such as the deletion of Spender's explicit line "I haven't any faith in humans," the effect of this is to enrich Spender and to humanize him;
d. it puts at the forefront the parallel with the European colonization of the Americas, a strong theme developed throughout the *Chronicles*.

Among smaller changes are slight shifts of emphasis in Bradbury's descriptions. In the magazine version we learn a little about the Martian towns:

> Really, it looked to all of them, as if the Martians were a tribal or family lot, one or another of the families from one town would find a green spot in the hills and a villa would be built with a pool and a library and some sort of stage and a good many balustrades and tiled terraces [*Thrilling Wonder Stories* 86].

In the *Chronicles* the above passage is replaced with this:

> Above the towns, scattered like pebbles, were single villas where ancient families had found a brook, a green spot, and laid out a tile pool, a library and a court with a pulsing fountain. Spender took half an hour, swimming in one of the pools which was filled with seasonal rain, waiting for the pursuers to catch up with him [*Silver Locusts* 83].

In this one short passage, Bradbury has shifted from the crew's interpretation of what they see ("it looked to all of them") to a statement of what the Martians had done, linked it with an active simile ("scattered like pebbles") and provided a tactile interaction between Spender and the Martian landscape.

Shortly after this scene, Spender finds a Martian book. In the magazine version of the story the book has an aluminum cover, "stamped" in black and gold. In the *Chronicles* it is a "silver" book, "hand-painted," signifying an increase in value and craftsmanship—and an increased significance to the visual and tactile allure of the object. These small stylistic adjustments alter the connectedness between Spender and Mars, and between the reader and Mars.

Bradbury's Allure: Style?

Certain of Bradbury's key stories have been adapted again and again into many different media. Among the most adapted of his stories are "The Veldt," "A Sound of Thunder," "Mars Is Heaven!," "And the Moon" and "There Will Come Soft Rains." What they all have in common is a fantastical element disrupting the commonplace ... but this same observation is true of many other Bradbury stories, and of stories of many other writers. There must be something more.

Many critics have identified Bradbury's style as his unique quality, and the stories listed above all exhibit his hallmark vivid descriptions and use of metaphor. Eller and Touponce refer to Bradbury simply as a "short story writer with a characteristic 'poetic' style" (xvii). Mengeling (885) identifies a distinctive combination of concrete and abstract terms, and what he calls an "intense concentration," in the construction of Bradbury's descriptive passages. McGiveron (177) identifies a number of stylistic devices used commonly in Bradbury's writing. Pell (186–194) catalogues Bradbury's devices across three book-length works, and determines metaphoric style to be the key to his work.

As Eller and Touponce have pointed out, though, Bradbury actually switches quite ably between the metonymic and metaphoric poles identified by Roman Jakobson. He tends to the metonymic pole when clarity of action is important, swinging to the metaphoric at precisely those moments when the fantastic must disrupt the commonplace. His metaphoric flourishes are never so powerful as when a T. Rex puts in its first appearance ("A Sound of Thunder") or when lions come to life in a virtual reality nursery ("The Veldt").

"And the Moon" occupies a peculiar position in this regard. The metaphorical flourishes are few, and most detectable when describing a passive landscape. I would argue that the allure of "And the Moon" lies only partly with the outwardly stylistic aspects. Greater attractions, for would-be adapters to other media at any rate, must surely be the intensity of the interior monologue which narrates the earlier sections of the story, and the clear dramatic conflict which stands between Spender and Wilder.

While "And the Moon" is told by an omniscient narrator, for key passages we enter into the mind of Spender. This is an effective device for engendering sympathy for an ultimately destructive character, and the motivational adjustments made in the *Chronicles* version of the story only help to strengthen our sympathy for him. The classic dramatic triangle then represented by Spender-Wilder-Biggs allows us to explore a moral spectrum of views relating to the rights and wrongs of colonization: Spender is against it, but all too keen on immersing himself in the "true" Mars; Biggs is all in favor of ripping the planet apart and remaking Mars in his own image; Wilder is at the centre, forced to make a decision based on informed pragmatism.

"And the Moon" is one of the few chapters of *The Martian Chronicles* to have such a sharply defined cast of characters. Spender and Wilder meet all of the criteria for good dramatic characters identified in screenwriting bibles such as that by Parker, so of all the chapters in the *Chronicles*, it is not surprising that this one has become so often dramatized.

Adaptations by Others: Radio Successes, TV Struggles

"And the Moon" was first adapted for radio's *Dimension X* in 1950, by Ernest Kinoy. Kinoy extrapolates Bradbury's description of the dead Martians, describing the effect of chicken pox as "burned them black and dried them out to brittle flakes."[3] Beyond this slight enhancement of Bradbury's metaphor, Kinoy presents a straightforward adaptation of the Martian landscape.

Perhaps the best-known adaptation of "And the Moon" is within the 1980 television miniseries *Ray Bradbury's The Martian Chronicles*, scripted by Richard Matheson and directed by Michael Anderson. As with other (unfilmed) attempts at full-length adaptation of the *Chronicles*, Matheson adopts a strategy of maximizing continuity in the adaptation by recasting certain characters so that they reappear in other episodes, a strategy Bradbury had already adopted to an extent in his fixing-up[4] of *The Martian Chronicles*. Spender appears as a minor character long before the "And the Moon" episode, notably at a party where he debates the colonization with Wilder's wife. In this version, Spender is the *only* crewman to survey the dead Martian towns.

Television, of course, opens up the opportunity for visually representing the Martians and their civilization. On an architectural level, this production does so magnificently. The Martian city is all simple shapes: spheres, columns, pyramids, crystalline forms: we are reminded of the "fragile towers" of the original short story (*Thrilling Wonder Stories* 91).

What this production lacks, however, is any interaction between the characters and the Martians' world. We never go inside any of the Martian dwellings, never see any of the books Bradbury refers to, never see Spender swimming in a Martian pool. The sense of a lived landscape is lost.

One of the few attempts to connect character to landscape comes through Spender. When he kills, it is with a Martian weapon. More significantly, when Spender is hunted down and killed he is found — without explicit explanation — dressed as a Martian, and wearing a Martian mask.

Masks have long been used as a dramatic symbol of a change or a hiding of an identity, and they feature in *The Martian Chronicles*—but not in Bradbury's text for "And the Moon." As we shall see, the adoption of a mask symbol is significant in Bradbury's own media adaptations of the *Chronicles*.

Difficulties of Adaptation

The TV miniseries of *Chronicles* is widely seen as a dramatic failure. Bradbury himself dismisses it as "boring" (qtd. in Szalay 38). With the debatable exception of Truffaut's *Fahrenheit 451*, most screen adaptations of Bradbury's work (to this point, at least) have been similarly dismissed. Indeed, it is a popular contention that Bradbury just doesn't adapt well. Rod Serling said it: "Bradbury is a very difficult guy to dramatize because that which reads so beautifully on the printed page doesn't fit in the mouth — it fits in the head" (qtd. in Zicree 274). Matheson says it (Archive of American TV interview): "His prose is like poetry ... and to get that on the screen, in the harsh realistic terms of the screen, is almost impossible."

But with so many different methods of adaptation available to writers and directors, why should this be? Elliott (220–243) has identified a half-dozen strategies used by adaptors, and found viable adaptations associated with each one. From Elliott's schema, we might recognize the "genetic" concept of adaptation in the case of *Dimension X*: Something in the underlying structure of the source text informs the development and growth of the offspring drama. We might also recognize the "incarnational" concept in the case of *The Martian Chronicles* miniseries: The series appears for all the world to be Bradbury's book made flesh.

More likely for the miniseries is that it is a failed example of Elliott's "psychic" category: The novel's "spirit" is expected to somehow infuse the filmmaker, but actually fails to do so. I would argue that this is in part due to a misapprehension of the allure of the source text. Bradbury's style draws the reader into his narrative, making the text attractive. But the adaptation to another medium is unable to use Bradbury's signifiers directly, leaving an uninspired shell of his signifieds; this is what Matheson is acknowledging. To use Mengeling's terminology, from the concrete-abstract pairing, the abstract does not translate too easily, leaving only the concrete. To use Jakobson's terminology, if the metaphoric pole cannot be realized in an adaptation, all that remains is the metonymic.

Another view is suggested by Harlan Ellison, in pondering the failure of adaptations of some very popular novelists. He places Bradbury in company with Pinter, Hemingway and Stephen King as "profoundly allegorical writers" who "*seem* to be mimetic, but they *aren't*" (181). "And the Moon" is one of the most allegorical tales in the *Chronicles*.

Self-Adaptation and the Controlling Influence

Bradbury's own adaptations show a willingness to interrogate his original texts, something that recent scholarship has revealed to be a fundamental characteristic of his working methods: Eller and Touponce show that Bradbury is an inveterate rewriter.

Most of the literature on adaptation has little to say of the self-adapting artist.[5] Bradbury is extensively self-adapting. He has adapted most of his major novels for other media (*The Martian Chronicles*, *Fahrenheit 451*, *Something Wicked This Way Comes*, *Dandelion Wine*), and many of his works in print are actually adaptations of media originals (the novellas "Leviathan '99" and "Somewhere a Band Is Playing"; *Something Wicked This Way Comes*; the short story "I Sing the Body Electric"). Uniquely among major authors, he also adapted sixty-five of his stories for his own television series.

Bradbury has attempted adaptations of *The Martian Chronicles* for the screen on several occasions, and each time maintains "And the Moon" as a pivotal story.[6] He has also adapted *Chronicles* for the stage, using "And the Moon" as the story that opens Act Two, indicating its role as a major turning point in the overarching *Chronicles* narrative.

His 1961 screenplay, written for MGM, introduces Spender on page 2, acknowledging his significance to the overall *Chronicles* arc. Spender makes a distinction between what men do (build rockets) and what nature does (provides the rocket's fire). This may be a thematic "excuse" for the colonization of Mars — we send the rocket ships, but nature causes the Martians to die; or a thematic foreshadowing of his own fate: he propels himself into the Martian city, but nature causes the Martian to take him over and become a killer.

In Bradbury's 1964 screenplay, Spender's death has him entangled in a long, unraveling red ribbon or sash (Nolan). The red sash is a red gash; blood laid out across the landscape of the Red Planet. The other crewmen hover around on their jetpacks, and give a striking visual of an airborne funeral cortège; they could be angels. Bradbury is clearly not constrained by the "stark realism" that Matheson refers to, suggesting a fundamentally different approach to adaptation.

In Bradbury's 1978 teleplay, Bradbury uses the voices and masks of Mars as multiple narrators, an echo of an idea developed in his stage play. The landscape is framed partly through the vision of the Martians themselves.

The textual variations I have catalogued so far might easily be interpreted as an exploration of medium-specificity (Hutcheon 33–34). However, Bradbury's adaptations of "And the Moon" demonstrate an elaboration and extension of certain metaphors that have a powerful impact on the depiction of the Martian landscape, and of the characters' relationships with it.

One of the clearest instances of this is his development of the "dead leaves" that blow around the surface of Mars. The leaves are actually the desiccated remnants of the last Martians. They make no appearance at all in the original magazine short story. In the *Chronicles* chapter, Hathaway's account of the Martian town mentions "walking in a pile of autumn leaves. Like sticks and pieces of burnt newspaper," implying them to be the remains of Martians. By the time of Bradbury's stage play — first performed in 1976 (Weist), and published in 1986 — the *black* leaves are everywhere in the "And the Moon" section, including at the landing site, which is where the crew does an analysis, concluding: "Here they [earthly bacteria and viruses] burn you to ashes—flake you away like charcoal and dust." There are echoes here of other Bradbury works: *Fahrenheit 451* ("we burn them to ashes, then burn the ashes") and, when the dust is compared to the dust from a mummy, *From the Dust Returned*.

In developing the "leaves" metaphor, Bradbury is both extending and darkening what is signified. Curiously, he is also shifting from an effect of evocation to an attempted objectification of the death of the Martians. Just as his insertion of the discovery of recently dead Martians provides Spender with an objective basis for his motivations, so the scientific analysis of the leaves provides "proof" of the presence of Martian DNA.

Another continuing line of development is Spender's engagement with Martian technology and artifacts, and his consequent role in the Martian landscape. In the stage play, Spender is introduced as one of the crew of "The Third Expedition," so that when Wilder's fourth expedition arrives, it finds Spender already present, a fixture in the Martian landscape. Spender explicitly declares, "My name isn't Spender any more," announcing a distinct change of identity which in previous texts has simply been implied.

For much of the play, Spender carries a Martian mask, and when Cherokee[7] urges him to remove it, he finds that he cannot do so. Later, he reports, "I hadn't *changed* enough and become completely Martian"; and then, "I found my madness," as if his state of mind has a geographical component.

A further elaboration of Spender's identity comes in a telepathic link between Spender and the captain. Though always intimately linked in all of Bradbury's texts, in the play the two characters and the landscape become highly linked in a literal manner.[8]

In 1990, Bradbury adapted "And the Moon" for *The Ray Bradbury Theater* as a stand-alone story. The teleplay, which appears to survive only in a second draft manuscript, differs significantly from the episode as filmed. Bradbury evidently produced a later draft which no longer survives. Comparison of the second draft with the filmed episode reveals a tension between two controlling influences: Bradbury's variant text, developed across multiple screen and stage adaptations, and the original story. This results in a struggle between three of Elliott's concepts of adaptation.

Should the adaptation be "genetic," with a clear inheritance of ideas between source text and media adaptation? Should it be "incarnational," with the TV episode becoming the embodiment of the "classic" text? Or should the adaptation "trump" the source text, and prove to have values above and beyond its parent?

Bradbury's method for his teleplay adaptations was to write without looking back at the original versions of his short stories. Only after drafting the teleplay would he return to the text to see if he had missed anything: "this gives you the intellectual leeway to do things that improve the story.... I have respect for my younger self, but I don't let that override my ability, 40 years later, to improve it" (qtd. in Warren 30).

In the episode as filmed, Spender believes he has found some kinship with the extinct Martians. He inhabits a Martian dwelling, engages with their artifacts, dons a mask. More so than in Bradbury's original story, Spender is echoing Krutch, who wrote:

When the life has entirely gone out of a work of art come down to us from the past, when we read it without any emotional comprehension whatsoever ... it has ceased to be a work of art at all and has dwindled into one of those deceptive "documents" from which we get a false sense of comprehending through the intellect things which cannot be comprehended at all except by means of a kinship of feeling [79–80].

At the end of the draft teleplay, Spender is entombed, and Wilder places the mask on him. Wilder acknowledges his inherited role as the conscience of Mars. In the episode as filmed, Spender gives a mask to the captain, telling him "You're Spender now." The captain dons the mask, and the implication is that he will inherit Spender's madness/delusion. Free of the constraints of *The Martian Chronicles* continuity, this stand-alone episode is able to adopt a more ominous tone than the original story.

The tension surrounding the ultimate controlling influence for this adaptation is resolved by the adoption of Elliott's remaining category of adaptation, "de(re)composing," in which the adaptation becomes a composite of signifiers and signified from the film and all previous adaptations, although what is hidden from the viewer is the wealth of unpublished drafts which weave even more variants of the story. Bazin proposes that if an original text is like a crystal chandelier, a film adaptation is a flashlight which intersects and illuminates it (Andrew 35). Bradbury's self-adaptations are somewhat more complex, being simultaneously the flashlight and an elaboration of the chandelier.

Conclusions

Bradbury's stories have an allure which seems impossible to capture in a literal adaptation. Adapters come unstuck when they attempt a prosaic, mimetic adaptation of his work. "And the Moon Be Still as Bright" is a rare combination of small bursts of metaphor set in an otherwise character-driven, metonymic narrative, which has allowed some adaptational success, particularly in radio.

But the author himself seems free of the controlling influence of his source text when self-adapting, and in his freewheeling adaptations has achieved extended metaphors for the relationships between humans and Mars, as manifested in the "human space" of occupied Mars. It is only when the desire to provide an "incarnational" adaptation interferes that a struggle for controlling influence ensues, but with the result that a hybrid media text emerges, which both extends and illuminates the original.

A study of Bradbury's self-adaptation is long overdue. Now that more of his media work is coming into print, it is becoming possible to see how consideration of alternative forms has influenced his developing authorship.

Notes

1. A persuasive account of Bradbury's use of nostalgic imagery in terms of Turner's American frontier thesis is given by Wolfe. However, some of Wolfe's assumptions need to be revisited in light of recent scholarship on the textual history of The Martian Chronicles and related stories.

For example, Wolfe assumes "Dark They Were and Golden Eyed" was left out of *The Martian Chronicles* because it doesn't come from the same chronology — but in fact Bradbury did have a place for this story in earlier drafts of his table of contents. Similarly, the assumed "obvious" ending of the book ("The Million-Year Picnic") was, for a long time in Bradbury's plans, intended to be a long way from the conclusion of the volume. Eller and Touponce (2002) give the most detailed textual history, and also summarize the key changes Bradbury made when he adapted the short stories into the novel of *The Martian Chronicles*.

2. I have been unable to locate any prior scholarship linking Bradbury and Krutch. However, Jon Eller is currently researching this area for his forthcoming book *Becoming Ray Bradbury*, which examines the intellectual and other influences on Bradbury's authorship.

3. This line is excised from the surviving recording of the X Minus One episode, but is present in the surviving Dimension X.

4. "Fix-up" (as both noun and verb) is the term coined by A.E van Vogt for making a novel out of previously published short stories. A good account is given in an interview on *Icshi: the A.E. van Vogt Information Site* http://home.earthlink.net/~icshi/Interviews/Weinberg-1980.html.

5. Hutcheon gives one of the most systematic and widely-applicable accounts of adaptation, but makes almost no mention of adaptation by the original author of a text.

6. None of these adaptations has been filmed, and none have yet been published in full, although publication of some is expected in 2010 (Bradbury, Ray. *The Martian Chronicles: the Complete Edition*. Burton, MI: Subterranean Press, 2010 [expected date of publication]. Print.) I am indebted to the Center for Ray Bradbury Studies, Indiana University, for granting access to the unpublished manuscripts.

7. Bradbury's characters have suffered some awkward name changes through these extended processes of adaptation. The original short story and *Chronicles* chapter of "And the Moon..." both feature a character who starts off as Gibbs, but later becomes Biggs. Some later printings of *Chronicles* apparently correct this persistent mis-type, while Matheson/Anderson TV miniseries renames the character Briggs. Meanwhile, another mis-type leads Cheroke to be rendered as Cherokee in the stage play script.

8. A curious addition to the wealth of adaptations of "And the Moon..." is an audio drama version released in 1988 by *Omni Science Fiction-Science Fact* magazine. Produced by Mike McDonough, this production was something of a companion piece to the earlier NPR radio series Bradbury 13. The recording lacks any detail in the writing credits, and the producer recalls writing the adaptation himself. However, direct comparison with Bradbury's play reveals that this is actually a production of the play rather than the original short story.

Adaptations of Stories from *The Martian Chronicles*

Further details can be found on the author's website at www.bradburymedia.co.uk.

Rocket Summer/Prologue of stage play/As part of "The Martian Chronicles," episode of *Dimension X* (NBC radio 1950), re-made in *X Minus One* (NBC radio 1955)

Ylla/Act 1, Sc 1 of stage play/As part of "The Martian Chronicles," episode of *Dimension X* (NBC radio 1950), re-made in *X Minus One* (NBC radio 1955)/As part of *The Martian Chronicles* (NBC TV 1980)/As part of *75th Birthday Tribute to Ray Bradbury* (California Artists Radio Theater 1995)

The Summer Night/Act 1, Sc 2 of stage play

The Earth Men/Act 1, Sc 3 of stage play/*Escape* (CBS radio 1951)/*The Ray Bradbury Theater* (Atlantis TV 1992)

The Taxpayer/no adaptations to date

The Third Expedition [aka "Mars Is Heaven!"]/Act 1, Sc 5 of stage play/*Dimension X* (NBC radio 1950), re-made in *X Minus One* (NBC radio 1955)/*Escape* (CBS radio 1950)/*Think*/*ABC Radio Workshop* (ABC radio 1953)/*Future Tense* (WMUK radio 1976)/As part of *The Martian Chronicles* (NBC tv 1980)/*The Ray Bradbury Theater* (Atlantis tv 1990)

And the Moon Be Still as Bright/Act 2, Sc 1–3 of stage play/As part of "The Martian Chronicles," episode of *Dimension X* (NBC radio 1950), re-made in *X Minus One* (NBC radio 1955)/*Dimension X* (NBC radio 1950), re-made in *X Minus One* (NBC radio 1955)/As part of *The Martian Chronicles* (NBC TV 1980)/*Omni Audio Experience I* (Omni/Bonneville Media Communications 1988)/*The Ray Bradbury Theater* (Atlantis TV 1990)

The Settlers/As part of "The Martian Chronicles," episode of *Dimension X* (NBC radio 1950), re-made in *X Minus One* (NBC radio 1955)

The Green Morning/Act 2, Sc 4 of stage play

The Locusts/no adaptations to date

Night Meeting/As part of *The Martian Chronicles* (NBC TV 1980)

The Shore/As part of "The Martian Chronicles," episode of *Dimension X* (NBC radio 1950), re-made in *X Minus One* (NBC radio 1955)

The Fire Balloons [This story is not included in all editions of ***The Martian Chronicles***.]/Act 2, Sc 4 and 7 of stage play/As part of *The Martian Chronicles* (NBC TV 1980)

Interim/no adaptations to date

The Musicians/no adaptations to date

The Wilderness [This story is not included in all editions of ***The Martian Chronicles***.]/no adaptations to date

Way in the Middle of the Air/no adaptations to date

The Naming of Names/no adaptations to date

Usher II [This story is not included in all editions of ***The Martian Chronicles***.]/Act 2, Sc 5 of stage play/*The Ray Bradbury Theater* (Atlantis TV 1990)

The Old Ones/no adaptations to date

The Martian/*The Ray Bradbury Theater* (Atlantis TV 1992)/As part of *The Martian Chronicles* (NBC TV 1980)

The Luggage Store/no adaptations to date

The Off Season/Act 2, Sc 6 of stage play/As part of "The Martian Chronicles," episode of *Dimension X* (NBC radio 1950), re-made in *X Minus One* (NBC radio 1955)/As part of *The Martian Chronicles* (NBC TV 1980)

The Watchers/no adaptations to date

The Silent Towns/Act 2, Sc 7 of stage play/As part of *The Martian Chronicles* (NBC tv 1980)/*The Ray Bradbury Theater* (Atlantis TV 1992)/As part of *80th Birthday Tribute to Ray Bradbury* (California Artists Radio Theater 2000)

The Long Years [aka "Dwellers in Silence"] *Dimension X* (NBC radio 1951), re-made in *X Minus One* (NBC radio 1955)/*The Ray Bradbury Theater* (Atlantis TV 1990)

There Will Come Soft Rains/*Dimension X* (NBC radio 1950), re-made in *X Minus One* (NBC radio 1955)/As part of "The Martian Chronicles," episode of *Dimension X* (NBC radio 1950), re-made in *X Minus One* (NBC radio 1955)/Single play (BBC radio 1962)/Single play (BBC radio 1971)/Single play (BBC radio 1977)/Short film (Uzbek film 1984)

The Million/Year Picnic/Act 2, Sc 7 of stage play/As part of "The Martian Chronicles," episode of *Dimension X* (NBC radio 1950), re-made in *X Minus One* (NBC radio 1955)/As part of *The Martian Chronicles* (NBC TV 1980)

Works Cited

Andrew, Dudley. "Adaptation." *Film Adaptation*. Ed. James Naremore. New Brunswick, N.J.: Rutgers University Press, 2000.

"And the Moon Be Still As Bright." *Dimension X.* NBC Radio. September 29, 1950. Accessed February 2010. http://www.archive.org/details/OTRR_Dimension_X_Singles.

"And the Moon Be Still As Bright." *X Minus One.* NBC Radio. April 22, 1955. Accessed February 2010. http://www.archive.org/details/OTRR_X_Minus_One_Singles.

Bradbury, Ray. "And the Moon Be Still As Bright." *The Ray Bradbury Theater.* Platinum, 2005. DVD.

_____. "And the Moon Be Still As Bright." *Thrilling Wonder Stories* 32.2 (June 1948).

_____. "And the Moon Be Still As Bright" [unpublished teleplay]. *The Ray Bradbury Theater.* Atlantis Productions, 1990. Copy held by Center for Ray Bradbury Studies, Indiana University. Accessed August 2009.

_____. *Fahrenheit 451.* London: Panther, 1980.

_____. *From the Dust Returned.* New York: Avon, 2002.

_____. *It Came from Outer Space*, Colorado Springs, Colo.: Gauntlet Press, 2004.

_____. *The Martian Chronicles* [unpublished screenplay]. MGM Studios, 1961. Copy held by Center for Ray Bradbury Studies, Indiana University. Accessed August 2009.

_____. *The Martian Chronicles* ["additional materials/revision of six-hour script down to two-hour version for theatrical release"]. 1978. Copy held by Center for Ray Bradbury Studies, Indiana University. Accessed August 2009.

_____. *The Martian Chronicles* [script]. Woodstock, Ill.: Dramatic Publishing Company, 1986.

_____. *The Martian Chronicles* [unabridged reading by the author, with commentary]. Chivers Audio Books, 1986. Audiocassette.

_____. *The Silver Locusts* [UK title of *The Martian Chronicles*]. London: Panther, 1977.

Eller, Jonathan R., and William F. Touponce. *Ray Bradbury: The Life of Fiction.* Kent, Ohio: Kent State University Press, 2004.

Elliott, Kamilla. "Literary Film Adaptation and the Form/Content Dilemma." *Narrative Across Media: The Languages of Storytelling.* Ed. Marie-Laure Ryan. Lincoln: University of Nebraska Press, 2004.

Ellison, Harlan. *Harlan Ellison's Watching.* Novato, Ca.: Underwood Miller, 1989.

Grimsley, Juliet. "*The Martian Chronicles*: A Provocative Study." *The English Journal.* 59.9 (December 1970).

Hutcheon, Linda. *A Theory of Adaptation.* London: Routledge, 2006.

Krutch, Joseph Wood. *The Modern Temper.* San Diego, Ca.: Harcourt Brace Jovanovich, 1984.

Lomax, William. "Landscape and the Romantic Dilemma: Myth and Metaphor in Science Fiction Narrative." *Mindscapes: the Geographies of Imagined Worlds.* Ed. George Edgar Slusser and Eric S. Rabkin. Carbondale: Southern Illinois University Press, 1989.

"The Martian Chronicles." *Dimension X.* NBC. August 18, 1950. Radio. Accessed February 2010. http://www.archive.org/details/OTRR_Dimension_X_Singles.

Matheson, Richard. Interview for Archive of American Television, 2002. March 26, 2008. http://www.emmys.tv/foundation/archive/interviews.php.

McDonough, Mike. "Re: Bradbury 13." E-mail message to the author. January 28, 2010.

McGiveron, Rafeeq O. "Bradbury's Fahrenheit 451." *The Explicator* 54.3 (1996).

Nichols, Phil. "Adaptive Behaviours: Ray Bradbury's Short Fictions Re-Interpreted for Media." Science Fiction Across Media: Adaptation/Novelisation conference, University of Leuven, May 29, 2009.

Nolan, William F. *The Ray Bradbury Companion.* Detroit, Mich.: Gale Research, 1975.

Omni Audio Experience I. *Omni*, 1988. Audiocassette.

Parker, Philip. *The Art and Science of Screenwriting.* Bristol, U.K.: Intellect Books, 2006.

Pell, Sarah-Warner J. "Style Is the Man: Imagery in Bradbury's Fiction." *Ray Bradbury.* Ed. Martin H. Greenberg and Joseph D. Olander. Edinburgh: Paul Harris Publishing, 1980.

Ray Bradbury's The Martian Chronicles. Dir. Michael Anderson. MGM, 1980. DVD.

Rose, Mark. *Alien Encounters: Anatomy of Science Fiction.* Cambridge, Mass.: Harvard University Press, 1982.
Said, Edward W. "Invention, Memory, and Place." *Critical Inquiry* 26.2 (2000).
Schama, Simon. *Landscape and Memory.* London: HarperPerennial, 2004.
Szalay, Jeff. "*Starlog* Interview: Ray Bradbury." *Starlog* 53 (December 1981).
Warren, Bill. "At Play in the Business of Metaphors." *Starlog* 153 (April 1990).
Weist, Jerry. *Bradbury: An Illustrated Life.* New York: William Morrow, 2002.
Weller, Sam. *The Bradbury Chronicles.* New York: William Morrow, 2005.
Wolfe, Gary K. "The Frontier Myth in Ray Bradbury." *Ray Bradbury.* Ed. Martin H. Greenberg and Joseph D. Olander. Edinburgh: Paul Harris Publishing, 1980.
Zicree, Marc Scott. *The Twilight Zone Companion.* New York: Bantam, 1982.

Robert A. Heinlein and the Red Planet

David Clayton

1

"Once upon a time there was a Martian named Valentine Michael Smith" (3). So commences the novel hyperbolically described on the cover of a recent edition as "THE MOST FAMOUS SCIENCE FICTION NOVEL EVER WRITTEN." In 1961, such an opening would have been quite unconventional for a novel by one of the most respected writers of science fiction in the United States. First of all, the use of the classic formula for opening a fairy tale alerts readers to expect something other than an uncomplicated saga of exploits in outer space. More importantly, the insouciant tone indicates the passing of a genre. When material once treated quite seriously by writers — including Heinlein in his earlier career — can be introduced so lightly, it is an unmistakable sign the material has become outdated. But what is the genre in question? It is that of science fiction literature dealing with Mars.

This genre flourished in the period between the publication of two equally remarkable books, H.G. Wells's *The War of the Worlds* in 1898 and Ray Bradbury's *The Martian Chronicles* in 1950. Already an autumnal tone pervades a good deal of Bradbury's book. Here Mars is not the angry red planet that launched a nearly successful invasion of Earth in *The War of the Worlds*, but a landscape haunted by the ruins of the past. Some eleven years later, Mars could only be conjured up on paper by means of a self-consciously ironic gesture.

Actually, Heinlein used Mars as a setting for four works in all: *Red Planet* (1949), *Double Star* (1956), *Stranger in a Strange Land* (1961), and *Podkayne of Mars* (1963). These novels, although not connected by narrative links such as common action or characters, do not form a series so much as a cycle. *Red Planet*, a juvenile adventure novel, depicts how two human boys, Jim and Frank, living on Mars, foil the schemes of the monopolistic Company to make itself master of the planet. In *Double Star*, Mars only serves as a momentary backdrop for the adventures of Larry Smith, a small-time actor called in to impersonate a famous politician whose "role" he takes over completely when the latter dies. Similarly, in the last of the books, only a small part of the action takes place on Mars, after which the heroine and her uncle depart for Earth. A remark of hers made in passing might well suggest the book was Heinlein's adieu to the Red Planet: "...we Marsmen (not 'Martians,' please!— Martians are a non-human race, now almost extinct)..." (14).

Doubtless, *Stranger in a Strange Land* marks the high point of Heinlein's encounter with Mars and Martians, and, for that reason, it seems to me reasonable to focus on it as the most important of Heinlein's writings that use a Martian background. The emphasis he gives to the theme from the very first words of the novel directs our attention as much to it as to the protagonist — more precisely, it joins the two in a common strand. In addition, the passage cited above suggests that *Stranger in a Stranger Land* may have been an act of reckoning after which Heinlein had no interest in returning to the Red Planet. The qualification of the Martians as "now almost extinct" even contradicts what we find at the end of the previous work, in which the Martians, facing extinction, still possess the power of eradicating their neighbors in the solar system.

But we are not finished with that opening sentence. Heinlein is being less than honest with his readers. This "Martian" is in reality a "Marsman," the product of an adulterous relationship between two of the first explorers of Mars from Earth. After the relationship results in the killing of the persons involved, the offspring of this union is raised by the "real" Martians. When Smith, now in his twenties, is found by a subsequent group of explorers, he has little consciousness of his identity as a human being. In a certain but by no means accidental fashion, he resembles an interplanetary Natty Bumppo, biologically the child of one race, but culturally the inheritor of a quite different one. When he returns to Earth, the experience is as traumatic for him as it would have been for a young human to have been suddenly transported to Mars and thrust into a Martian nest.

Heinlein's Martians, although strange looking, are almost normal in comparison to Wells's. They are similarly characterized as possessing far greater intelligence than human beings, and as having a far older culture. In both these respects, they resemble Wells's Martians. Yet they are in no sense overtly belligerent. In both *Red Planet* and *Stranger in a Strange Land*, the Martians to a certain extent simply tolerate the human invaders that have landed on their planet with the amused patience of grandparents putting up with boisterous juveniles. But neither can they be characterized as benign. In *Stranger in a Strange Land*, we learn that they have totally destroyed one planet in the remote past, and, at the end of the novel, they are contemplating whether or not to subject Earth to the same fate. Once a Martian, always a Martian!

What seems important to notice at this point is that in choosing to use this narrative machinery — Mars and its indigenous inhabitants — Heinlein was drawing upon a science fiction tradition. Just as there had been a "Matter of Troy" and a "Matter of Rome" in medieval literature, there has been a "Matter of Mars" in the history of science fiction, dating at least from *The War of the Worlds*. No cosmic war occurs in *Stranger in a Strange*

Land, and most of the action takes place on Earth. Nevertheless, since the novel focuses upon the exploits of a supposed Martian — a fiction the author endorses in that opening sentence — are we not witnessing, however figuratively, a new invasion of Earth by the Red Planet?

Following in Wells's tracks, a host of pulp stories, comic books, and schlock movies featured Martian invaders. A high point of this figurative invasion of Earth by Martians happened in 1938, with Orson Welles's radio adaptation of the novel broadcast on October 30. The broadcast deserves our closer attention. The *succès de scandale* of the broadcast catapulted Welles into the headlines and landed him a contract with RKO that led to his making *Citizen Kane*. Yet it may have had other, less evident yet equally far-reaching consequences. The furor generated by the event could have suggested to an acute observer like Robert Heinlein that the genre had a far more promising commercial future than anyone had hitherto foreseen.

And we can add another, if less evident, link. Andrew Sarris has suggested in *The American Cinema: Directors and Directions 1929–1968* that *The Thing from Another World*, the 1951 alien-from-outer-space thriller, also shows the influence of the broadcast: "*The Thing*, with its understandable traces of producer Howard Hawks and its unearthly traces of uncredited Orson Welles..." (218). And it may have in turn given Heinlein an idea. At one point, after the Thing has begun wreaking havoc at the Arctic research station, the devious man of science Dr. Carrington — echoing Exodus 2:22 — tries to deter the clear-sighted Captain Hendry from making ragout out of the creature with the following words: "Captain, when you find what you're looking for, remember, it's a stranger in a strange land."

2

In 1586, the Elizabethan dramatist and pamphleteer Robert Greene (1558?–1592) brought out a work called *Planetomachia*— the Greek title could be glossed as "war of the worlds." A defense of astrology indebted to the writings of Giovanni Pontano and Antoine Mizauld, the book portrays a debate between the planetary deities Saturn, Jupiter, Mars, Sol, Venus, Mercurie, and Luna. Within this framework, three of the gods — Venus, Luna, and Jupiter — relate a tale called a "Tragedie," belittling the powers of a rival. Venus attacks Saturn, Luna Venus, and Jupiter Mars. Each "Tragedie" is prefaced by a sketch of the planetary deity and the qualities it produces. Mars, according to Jupiter

> is of a fierie and inflamed nature, annexed with such burning heate and hurtfull intemperancie, stirring in mens bodies such hotte and adust choller, and infusing qualities together with his irradiation, more apt to destroy, th[a]n to ingender or nourish ... so that the poets fitly have figured *Mars* to be the God of warres, and *Bellona* to be his sister [65–66].

Here we are in a world of pre-scientific mythological belief that had been kept alive, as Jean Seznec demonstrated in *The Survival of the Pagan Gods*, primarily through literary and artistic sources passed down from antiquity, if not without a considerable amount of distortion. Summarizing the centuries-long process that culminated in "the identification of the gods with astral bodies," Seznec comments that

> during the last centuries of paganism belief in the divinity of the heavenly bodies grew even stronger. The stars are alive: they have a recognized appearance, a sex, a character, which their names alone suffice to invoke. They are powerful and redoubtable beings, anxiously prayed to and interrogated, since it is they who inspire all human action [41].

Moreover, as Seznec goes on to relate,

> Increasingly — as, for example, among the Neoplatonists of the third century — obsession with the divine and the demoniacal began to mingle with concepts of natural law and mechanics. Science lost ground to superstition and magic, or at the very least became inextricably involved with them [42].

As Alexander Koyré tacitly makes clear in *From the Closed World to the Infinite Universe* (1957), the moment that decisively brought to an end this image of a universe inhabited by the gods came not with Galileo and his heliocentric hypothesis, but with Descartes, who "clearly and distinctly formulated principles of the new science" (99). In the Cartesian world picture, "There is nothing else ... but matter and motion; or, matter being identical with space or extension, there is nothing else but extension and motion" (101).

It would be harder to imagine a more drastic contrast to Greene's fanciful diversion than *The War of the Worlds*, which commences with a soberly scientific survey of what was known about Mars at that time. The Cartesian universe of extended substance has taken a further step in the direction of demythologization with the aid of Charles Darwin. Wells offers his readers a universe not merely devoid of pagan deities, but overshadowed by "an incessant struggle for existence" (8). Logically, we would expect the same to held true of *Stranger in a Strange Land*, written more than sixty years later, the product of a far more technologically advanced milieu than the late Victorian setting of Wells's novel.

Yet what do we discover in Chapter IX but an astrologer, a lady named Alexandra Vesant, née Becky Vesey — the assumed name is probably intended to evoke memories of Annie Besant, a famous Theosophist and follower of Madame Blavatsky. This stargazer is no back-alley psychic, but a private advisor to the scheming wife of the president of the World Federation, a United Nations–like organization that governs the world, and a figure of considerable influence, owing to the gullibility of her client. And what is Madame Vesant being called upon to do but cast the horoscope of Valentine Michael Smith.

Shades of the "Chaldean soothsayers, a fraternity as indispensable as it was illfamed" (Seznec 41). Antics like these suggest the period of late paganism more than the age of space exploration. It is as if Mars was reclaiming its astrological rights, but reclaiming them under the most unlikely circumstances. Michael's arrival on Earth seems to act as a catalyst for transporting its inhabitants back to the third century. Does he not in his final avatar metamorphose into one of those "powerful and redoubtable beings, anxiously prayed to and interrogated"?

3

In *Robert Heinlein: America as Science Fiction*, H. Bruce Franklin offers a significant insight into his subject:

> Because he embodies the contradictions that have been developing in our society ever since the Depression flowed into the Second World War, to understand the phenomenon of Robert Heinlein is to understand the culture that is the matrix for ourselves [6].

As the subtitle of Franklin's book emphasizes, Heinlein's literary work is "America as Science Fiction." In the course of his survey, Franklin does a formidable job of showing how all the major conflicts of twentieth-century American history find their way into Heinlein's pages, viewed from a perspective of science fiction.

I would like to somewhat reformulate Franklin's thesis. If we think of history as a canvas a writer may inevitably recur to in fashioning an imaginative world, then Heinlein is probably one of the most important historical novelists to have appeared in America since James Fenimore Cooper — and the parallel will turn out to be not at all fortuitous. But Heinlein views history from a radically different perspective. Cooper launched the Leatherstocking tales with *The Pioneers* (1823), employing a more or less contemporaneous setting, yet in his subsequent novels he went farther back into the past, in a quest to discover the roots of the newly born American republic. Heinlein, by contrast, is always projecting the history he was living in the present into a possible future — no mean feat.

Heinlein's point of view is often more complex and subtle than Cooper's. On the one hand, contemporary events powerfully resonate throughout all of Heinlein's writing, from the earliest period down to his last published writings. It is one of Franklin's achievements to have detected these traces, like seismic tremors, and to have brought them into the light of day. On the other hand, Heinlein took his calling as a science fiction writer seriously from the beginning. If present-day occurrences — for example, the Cold War, in many novels of the 1950s and 1960s — could supply materials for a science fiction scenario, they still had to be displaced into the future, and the writer had to anticipate in what ways — first and foremost, technologically — a future world would reshape that scenario.

4

Although events in contemporary American history played a powerful role in shaping Heinlein's writing throughout his career, their influence is particularly conspicuous in the group of novels that appeared in the 1960s. Franklin has no difficulty in demonstrating the impact of the Cold War, of the rivalry between East and West, and of the fear of imminent nuclear war, on Heinlein's literary efforts during this period, but especially in *Starship Troopers* and *Farnham's Freehold*, in which a nuclear attack serves to precipitate the book's action. By contrast, *Stranger in a Strange Land* occupies a place by itself. Here Heinlein has moved the story only into a more or less proximate future, one almost indistinguishable from the present. Nevertheless, once the book finally found its readers, it achieved a resonance hardly equaled by any of Heinlein's other successes.

As Franklin recounts in the section of his monograph devoted to *Stranger in a Strange Land*:

> Unlike his prewar and postwar short stories, juvenile space epic, and hard-core science-fiction novels of the 1950s, *Stranger* did not immediately find its audience.... But in the mid–1960s, as millions of young people began to rebel against the most sacred American values of the 1950s, the novel became widely adopted within the movement known as "the counterculture," and its influence began bubbling around in the powerful, if confused, underground eddies and streams seething beneath what seemed the concrete and steel permanence of America's world of business [126–27].

Intentionally or not, Heinlein would seem to have made good something a few science fiction writers have dreamed of since the days of Hugo Gernsback: to make predictions about the future that turn out to be fulfilled in fact. At the time Heinlein wrote, the most dramatic manifestations of "the counterculture," such as the politicization of college students, mass demonstrations against the Vietnam War, and the widespread use of mind-altering substances, all had yet to happen.

The impetus for *Stranger in a Strange Land* came from *The Red Planet*, but in the long period between *Stranger in a Strange Land*'s inception and its ultimate appearance in print, Heinlein's ideas would have undergone a considerable period of incubation. At some point, he must have decided the main action would concern a "Martian" who brings a new religion to Earth; the title itself echoes Exodus 2:22 and hints at a parallel between Mike and Moses. But then Heinlein would have needed to furnish his protagonist with some kind of remotely credible doctrine to expound. Here is how Franklin summarizes Smith's teachings:

> Heinlein takes the logic of Michael's extreme philosophic communism all the way to an attempted breakout from his most threatening predicament — the narcissism and solipsism that leaves the isolated "free" individual in cosmic loneliness. Dr. Mahmoud, the Muslim "semantician" from the spaceship that had found Michael on Mars, explains the philosophic significance of the verb "grok": "The Martians seem to know instinctively what we learned painfully from modern physics, that observer interacts with observed through the process of observation. 'Grok' means to understand so thoroughly that the observer becomes part of the observed — to merge, blend, intermarry, lose identity in group experience" (Ch. 21). Mike tries to sum this up in his customary admonition, "Thou art God," apparently the closest expression to a Martian word that means "the universe proclaiming its self-awareness" (Ch. 31). Ben explains to Jubal that Mike believes "that whenever you encounter any other grokking thing — man, woman, or stray cat ... you are meeting your 'other end.' The universe is a thing we whipped up among us and agreed to forget the gag" (Ch. 31) [139–40].

A phrase like "narcissism and solipsism," which Franklin launches at Heinlein on more than one occasion, makes for decoratively pyrotechnic polemics, but it doesn't give us any insight into what might have led the author to pen these lines. First of all, the sentiments are by no means very original, either with the fictional Smith or his real-life creator. In spite of the Hegelian sound of "the universe proclaiming its self-awareness," no one should have any difficulty in detecting an updated version of the myth of what R.W.B. Lewis in a famous study published in 1959 called the "American Adam," the idea that the United States marks a radically new chapter in world history, properly speaking a "re-beginning," a return to the unlimited potential for human development that had been lost through Adam's fall.

This myth, although it finds unmistakable expression in Ralph Waldo Emerson's early essays such as "Nature," has deep roots, going back to colonial days. In effect the myth is an inverted Calvinism, transforming mankind's negative distance from God into a positive quantity. The human race's history is no longer perceived as a progressive descent into greater and greater depravity, but rather as a long, but necessary peregrination, at the end of which mankind would recover its pre-lapsarian powers. Ironically, Calvinism, by so dramatically emphasizing sinning mankind's estrangement from God, turned out to be the most powerful instrument for producing its own diametrical opposite. In that sense, Franklin hardly errs when he writes that "Michael's religion ... reveals a core of the most fundamental brand of Calvinism" (136).

But he seems to me to underrate the implications of his own insight. There is more at work than "an elite capable of being saved." Here is how Heinlein depicts Michael at the moment he reveals himself to the mob that will soon destroy him:

> His clothes vanished. He stood before them, a golden youth, clothed only in beauty — beauty that made Jubal's heart ache, thinking that Michelangelo in his ancient years would have climbed down from his high scaffolding to record it for generations unborn. Mike said gently, "Look at me. I am a son of man" [428].

Whatever else he might be, this figure is no meekly suffering Jesus, but the American Adam reincarnated as science fiction hero. Presented in this way, Michael is Nietzsche's Antichrist, the triumph of paganism over Christianity. Presumably Michelangelo would have abandoned painting the creation of man by Jehovah on the Sistine Chapel ceiling to paint Michael Valentine Smith's new creation.

5

Viewed coldly, outside of the heady atmosphere of the late 1960s, the concluding section of *Stranger in a Strange Land*, "His Happy Destiny," seems utterly equivocal, as much satire as hagiography. The former reaches its high point in the scene of Michael's martyrdom. Here is how the text continues after the rhapsodic description cited above:

> The scene cut for a ten-second plug, a line of can-can dancers singing:
> "*Come* on, *la*dies, *do* your *duds!*
> *In* the *smoo*thest, *yummi*est suds!
> *Lov*er *Soap* is *kind* to *hands* —
> *But* be *sure* you *save* the *hands!*"
> The tank filled with foamy suds amid girlish laughter and the scene cut back to the newscast:
> "God damn you!" a half brick caught Mike in the ribs. He turned his face toward his assailant. "But you yourself are God. You can damn only yourself ... and you can never escape yourself."

> "Blasphemer!" A rock caught him over his left eye and blood welled forth.
> ...
> Through bruised and bleeding lips he smiled at them, looking straight into the camera with an expression of yearning tenderness on his face. Some trick of sunlight and stereo formed a golden halo back of his head. "Oh my brothers, I love you so! Drink deep. Share and grow closer without end. Thou art God."
> Jubal whispered it back to him. The scene made a five-second cut: "*Cahuenga Cave!* The nightclub with real Los Angeles smog, imported fresh every day. Six exotic dancers."
> "Lynch him! Give the bastard a nigger necktie!" A heavy gauge shotgun blasted at close range and Mike's right arm was struck off at the elbow and fell. It floated gently down, then came to rest on the cool grasses, its hand curved open in invitation [428–429].

The effect of Heinlein's bravura style here is to render his hero's apotheosis utterly void. We need only compare this episode with a justly famous immolation scene, the slaying of Joe Christmas in William Faulkner's *Light in August,* to see the difference. Christmas is more a pathetic victim than a tragic hero, but at that moment he transcends his sordid destiny. A solemn aura surrounds the martyrdom of Christmas, but Mike's is turned into an infernal carnival in which some of the most terrifying aspects of American culture, mob violence and the degradation of life to a common denominator of commercial exploitation, fuse together. Mike gets his fifteen minutes of fame, and it's an apotheosis of nihilism.

By depicting Mike's martyrdom from Jubal's point of view as he watches it on television, Heinlein transforms it into a "media event" that robs Smith's Passion of any hint of transcendence. Even the halo that mystically appears is characterized as "a trick of sunlight and stereo," an image that suggests obscene kitsch like paintings of the crucifixion in day-glo colors on black velvet. In many ways, this "anti-transcendence" has been anticipated by the long, rambling, and not especially lucid exchange between Smith and Jubal in Chapter XXXVI.

The shortcomings of this episode notwithstanding, we can sort out three related strains: Smith's mission, his relation to the Martians, and the future of his cult. First, Smith admits — as we know already — he was used as a spy/missionary by the Martians:

> [The Old Ones] linked with me but left me on my own, ignored me — then triggered me, and all I had seen and heard and done and felt and grokked poured out and into their records. I don't mean that they wiped my mind of it; they simply played the tape, so to speak, made a copy. But the triggering I could feel — and it was over before I could stop it. Then they cut off the linkage; I couldn't even protest [416].

The language used, the emphasis on Smith's passivity, his inability to resist the Martians ("I couldn't even protest"), coupled with phallic imagery that suggests manipulation ("triggered me"), followed by ejaculation ("all I had seen and heard and done and felt and grokked poured out and into their records") — all these features suggest a full-blown paranoid fantasy like that of Daniel Paul Schreber, butt-fucked by God with the plan of transforming him into a woman, as detailed in his *Memoirs of My Nervous Illness* (*Denkwürdigkeiten eines Nervenkranken* [1903]).

More than anything else, *Stranger in a Strange Land* is a classic American fable about loss of innocence — a point I am surprised Franklin, who has an excellent knowledge of American letters, overlooked. The most interesting of the three strains that run through Smith's interview with Harshaw is that of the former's own realization of his loss of innocence and his ensuing disillusionment. Heinlein makes the point clearly through the topological organization of the novel, with the division of the setting into two discrete spaces, Earth and Mars, each of which has its own respective sphere of action. But Mars takes precedence, just as it did in ancient astrology. Populated by a race of super-beings, the planet might as well be an incarnate god back in the third century. When Smith goes to Earth, it is just as if he is falling from the heavens, with their ineffaceable purity, into a cesspool — and it fatally contaminates him. In more than one way, Smith is an "angel" — etymologically the word means "messenger" — but he forfeits his angelic status by his stay on Earth.

It is not only the topological distance that counts. There is also a compositional distance lying between Heinlein's first venture on Mars in *The Red Planet* and his most celebrated one, eleven years later. That distance is also a historical one, the time traversed both by Heinlein and the United States between 1949 and 1961. If we were to assume James Madison Marlowe, Jr., to be thirteen years old when *The Red Planet* takes place, he would be exactly Valentine Michael Smith's age when the latter first arrives on Earth. But what a change has occurred! The tone of the 1949 book is exactly as Franklin describes it: "optimistic, expansionary, romantic." The corrupt civilization back on Terra is merely hinted at by the machinations of the Mars Company, an interstellar equivalent of the British East India Company, as Franklin notes (79). But terrestrial decadence is completely eclipsed by the healthiness of frontier life on Mars. What we find in the later book is quite a different matter.

If we try to imagine Jim's quite strait-laced, stereotypically conventional family engaging in group orgies with their next-door neighbors, then we have a pretty concrete image of what separates the two books. Although Heinlein charmingly allows the possibility of an adolescent homosexual attraction in the friendship between Jim and his buddy — this is the frontier, after all — Jim is an innocent, and not necessarily in the sense of being sexually inexperienced. Smith's loss of innocence comes not through his promiscuity but through his seduction by power. If he wants to cure the "diseased and crippled" humans, how can he do so except by first infecting himself?

Yet in true Puritan fashion Smith has to rationalize his self-infection as an obligation imposed on him by his divine mission:

> He went on to Jubal, "'Thou art God.' It's not a message of cheer and hope, Jubal. It's defiance — and an unafraid unabashed assumption of personal responsibility." He looked sad. "But I rarely put it over. A very few, just these few here with us, our brothers, understood me and accepted the bitter along with the sweet, stood up and drank it — grokked it. The others, hundreds and thousands of others, either insisted on treating it as a prize without a contest — a 'conversion' — or ignored it. No matter what I said they insisted on thinking of God as something outside themselves. Something that yearns to take every indolent moron to His breast and comfort him. The notion that the effort has to be *their own* ... and that the trouble they are in is all their own doing ... is one that they can't or won't entertain."

The Man from Mars shook his head. "My failures so greatly outnumber my successes that I wonder if full grokking will show that I am on the wrong track — that this race must be split up, hating each other, fighting, constantly unhappy and at war even with their own individual selves ... simply to have that weeding out that every race must have" [423].

It is as if we are witnessing a return of the Calvinist repressed. Instead of an affirmation of life, we have the call to a strict, life-denying code of morality. Does Smith believe it himself? Or does he, in any case, believe himself capable of teaching it to others?

A profound ethical nihilism underlies Smith's efforts. Already the use of a con game, peddling his teachings as religion, makes his task problematic. He passes off this question in a nearly parenthetical aside as an "ends-justifies-the-means" detail of secondary importance: "What I had to teach couldn't be taught in schools; I was forced to smuggle it in as a religion — which it is not — and con the marks into tasting it by appealing to their curiosity" (419). That "which it is not" is crucial. If Smith's teaching is not religion but rational belief, then what could justify deceiving its prospective followers? What right does he have to complain his followers don't understand his teachings when he has led them astray from the first word out of his mouth?

Heinlein's Earth is a doomed planet, a point he made explicit a few years later in *Time Enough for Love* (1973): "The death rattle of Earth was clear and strong back in the twentieth century" (Franklin 181). We have already noted the proliferation of con men and deceitful cults that goes hand in hand with the pervasive gullibility of Earth's inhabitants. From the beginning, Heinlein depicts this future of the human race as a nightmare vision of consumerism and collectivism run amok. But the appropriate genre for such a fallen world is satire, not the tragic death of a god or the Christian "Divine Comedy" of a resurrected one. Paradoxically, satire, not messianic vision, is *Stranger in a Strange Land*'s strong card. In this latter-day "Planetomachia," Robert Greene's "tragedie" turns into a satyr play.

6

Perhaps Heinlein's turning away from Mars after *Stranger in a Strange Land* represents an abandonment of hope for Earth and its beings as well. But does that imply the collapse of the species as well? Or will it succeed in finding a new frontier, a new beginning of the American Adam, in the depths of outer space? In his early career, Heinlein made use of an elaborate chronological scheme in sketching out a hypothetical "future history" of human civilization. As Franklin explains:

> Tacked on to the end of "Logic of Empire" is a note from editor John Campbell, informing readers of *Astounding* that "all of Robert Heinlein's stories are based on a common proposed future history of the world." Two months later, in May 1941, *Astounding* printed Heinlein's chart of this future history. Modeled on the charts of macrohistory included in Olaf Stapledon's *Last and First Men* (1930), and sharing Stapledon's vision of a spiral of progress moving upward through cyclical rises and falls, Heinlein's chart provided a framework for much of his prewar fiction, an independent

display of his historical ideology, and a new pleasure for his growing throng of readers, who could now anticipate the missing pieces of the puzzle [28].

As it appears in a paperback edition of the collection of early stories published under the title *The Past through Tomorrow*, the chart is divided by a series of columns labeled, respectively, "Dates," "Stories," "Technical," "Data," "Sociological," and "Remarks." The final entry in the last column reads as follows: "Civil disorder, followed by the end of human adolescence, and beginning of first mature culture" (661). The chart assigns two early stories, "Universe" (1941) and "Commonsense" (1941), to this period, but it is difficult to see that Heinlein — measured by his own standards — ever succeeded in giving his readers a convincing depiction of "the end of human adolescence," much less the "beginning of first mature culture." Certainly the saga of Lazarus Long in *Time Enough for Love* is not the answer, resembling as it does an adolescent fantasy in which the protagonist simultaneously triumphs over mortality and the second law of thermodynamics by recycling himself forever. In effect, Long symbolically achieves the male equivalent of parthenogenesis through a perpetual orgasm in which his repeated ejaculations enable him to fertilize and reproduce himself endlessly.

Even the boundless self-reliance of Emerson and Whitman never went quite so far. This scenario does not evoke a new beginning of society but the circular fatality of myth. Yet does not Heinlein give shape to this fatality by the figure of Mars, repeating its eternal trajectory? Franklin notwithstanding, I think it would go far to suspect that Heinlein was a closet adherent of astrology. On the other hand, by bringing such a highly charged image as that of the Red Planet into *Stranger in a Strange Land*, he was obliged to counter his vision of a perpetually reborn frontier with that of a contrary force of inertia — Nietzsche's "Human, all too human." Consciously or not, we remain fascinated by the myth projected into the astral beyond. We sit down here, grounded by our inability to get beyond the limits of our imagination, waiting for a Valentine Michael Smith to come to our rescue. And as long as we do, we remain on the threshold of "the end of human adolescence," without being able to cross it.

Works Cited

Franklin, H. Bruce. *Robert A. Heinlein: America as Science Fiction.* New York: Oxford University Press, 1980.
Greene, Robert. *Robert Greene's "Planetomachia."* 1585. Reprint, ed. Nandini Das. Aldershot, U.K.: Ashgate, 2007.
Hawks, Howard, prod. *The Thing from Another World.* Dir. Christian Nyby. Charles Lederer, adapt. By John W. Campbell, Jr. ("Who Goes There?"). RKO, 1951. DVD. Warner Home Video, 2003.
Heinlein, Robert A. *Double Star.* 1956. Reprint, New York: Del Rey Books, 1986.
_____. *The Past through Tomorrow.* 1967. Reprint, New York: Berkley Medallion Books, 1975.
_____. *Podkayne of Mars.* 1963. Reprint, New York: Ace Books, 2005.
_____. *Red Planet.* 1949. Reprint, New York: Del Rey Books, 2006.
_____. *Stranger in a Strange Land.* 1961. Reprint, New York: Ace Books, 1987.

Koyré, Alexandre. *From the Closed World to the Infinite Universe*. 1957. Reprint, Baltimore, MD: Johns Hopkins Paperbacks, 1968.

Sarris, Andrew. *The American Cinema: Directors and Directions 1929–1968*. New York: Dutton, 1968.

Schreber, Daniel Paul. *Memoirs of My Nervous Illness*. New York: New York Review of Books, 2000.

Seznec, Jean. *The Survival of the Pagan Gods: The Mythological Tradition and Its Place in Renaissance Humanism and Art*. 1940. Trans. Barbara F. Sessions. Reprint, New York: Harper Torchbooks, 1961.

Wells, H.G. *The War of the Worlds*. 1898. Reprint, ed. Patrick Parrinder. London: Penguin Classics, 2005.

BUSINESS AS USUAL
Philip K. Dick's Mars
Jorge Martins Rosa

The Land of (Lost) Opportunity

Few things set Philip K. Dick as clearly apart from golden age writers as his stance towards planetary colonization. While that "classical" attitude is usually euphoric, in Dick, as soon as *Solar Lottery*, not to mention earlier short stories such as "The Gun" or "Piper in the Woods," there is a thorough disappointment with the very possibility of leaving Earth and its natural satellite towards some other planet. While in *Solar Lottery* the tenth planet is nothing more than a wishful fantasy of an equivocal prophet, in later novels colonies frequently become inextricable from the narrative, but the pessimistic approach is still the rule, either in a planet as far as Whale's Mouth (in *The Unteleported Man*) or, more commonly, in Mars.

Our neighbor planet is indeed one of Dick's places of election for the establishment of a colony, particularly in the novels written in the sixties, so much that Carlo Pagetti, in "Dick and Meta-SF" (published in *Science-Fiction Studies* [*SFS*] no. 5), and later Kim Stanley Robinson, in his Ph.D. thesis on Dick, define part of the first half of that decade as the period of the "Martian Novels" (Pagetti 26; Robinson 51–64). Written in 1962 but published only in a modified form seven years later, *We Can Build You* helps to set the tone by portraying an overcrowded Earth where a real estate agent, Sam Barrows, has just "managed to get the United States Government to permit private speculation in land on the other planets ... Luna, Mars and Venus" (*WCBY* 31). Setting the trends of what would, even if seldom used, become a distinctive Dickian brand, the simulacra — that is, androids — produced by the Frauenzimmer family should be built having those future colonies (and colonists) in mind, "designed to look exactly like the family next door. A friendly, helpful family ... like you remember from your childhood back in Omaha, Nebraska" (*WCBY* 114), because "people are going to be lonely, there" (*WCBY* 115). Extraplanetary colonies are then still only a project for the near future, but we can already detect a commodified world in the making. Not an alternative to Earth but rather a place that resembles it barely enough to trick potential settlers into overcoming their reticence and heading towards that "new frontier."

Dick would return, notably in the masterpiece *Martian Time-Slip*, to that concept of Mars as "frontier." Wilderness, however, lies more in the process of colonization than in the planet itself. Living may be harsh for the settlers — who, by the way, do not have androids as "famnexdos" in the novel, as these only serve the purpose of teaching — and the natives may look primitive to Earth's standards, but "barbarism" is a word that should be applied only to those in power. As many Dickian scholars have noticed, in particular those gathered around *Science Fiction Studies*, Mars appears in that novel (and later, in *The Three Stigmata of Palmer Eldritch*) as nothing but a clumsy solution for the excess of population or some similar problem. Moreover, the Red Planet is a distorted mirror of our own while remaining an "ultraplanetary" (as in "ultramarine") province where corporations from the "metropolis" would be at large if it wasn't for the regulatory — but nevertheless ambiguous — role of the United Nations. If Dick's portrayals of Earth are already a caricature of his own social and political environment, when Mars comes to the front stage that attitude is intensified, to a point that Kim Stanley Robinson's claim that "Mars ... is a representation of the America in which Dick wrote the novels, in which certain facets of the society have been augmented, others suppressed" (Robinson 33) reads almost like a euphemism: the perpetual mending of old machinery and the parallel economy of smuggled goods evoke much more a nineteenth-century London slum à la Charles Dickens than Philip K. Dick's postwar America, described by Robinson as "an American suburb of 1963" (Robinson 55).

In *The Three Stigmata of Palmer Eldritch*, the other full-blown "Martian Novel,"[1] the economy of the colony may be more developed, but the only advantage of living on Mars seems to be a milder climate compared to the overheated Earth; other than that, life is yet again so unbearable that a whole industry is devoted to give the settlers an ersatz experience of an idyllic "normality"[2] through hallucinogenic drugs. But who can blame them? When Leo Bulero tries Palmer Eldritch's Chew-Z, there is almost a moral tone in his devaluation of the drug:

> This is all a hypnogogic, absolutely artificially induced pseudo-environment. We're not anywhere except where we started from; we're still at your demesne in Luna. Chew-Z doesn't create any new universe and you know it. There's no bona fide reincarnation with it. This is all just one big snow-job.... This is not even as real as Perky Pat, as the use of our own drug. And even that is open to the question as regards the validity of experience, its authenticity versus a purely hypnogogic or hallucinatory [*3SPE* 64].

It may be a twisted moral, but hardly more twisted than having to deal, back on Earth, with the possibility of being drafted to Mars if you are healthy enough.

In spite of those differences, either in *Martian Time-Slip* or *The Three Stigmata*'s Mars, there's no business like business. As Brian Aldiss argues in his short essay in *SFS* no. 5, the "web" is all-entangling. Carlo Pagetti, again in *SFS* no. 5, gives a more detailed (and also less euphemistic) account of that entanglement between economy and the individual:

> At a closer look, in fact, the planet of *Martian Time-Slip* is revealed as a replica of budding American society not only with its generous pioneers, but also with phenomena from the formation of a capitalist society dominated by the inexorable law of profit and speculation.... The values that dominate Martian reality are again the ruthless struggle for power: violence, deceit, and, finally, the spiritual aridity of man.... Mars is, therefore, another of the many images of the Waste Land that 20th century culture proposes to us with obsessive repetitiousness. If for T. S. Eliot ... history is a labyrinth without an exit, for the author of *Martian Time-Slip* the future is an incubus evoked by the mind of an autistic child, who projects into already nightmarish reality his terror of life and his inability to communicate with others [Pagetti 27].

Christopher Palmer, in an essay only indirectly related to the portrayal of planetary colonies, echoes Pagetti's words when speaking of a "sterility" that pervades those and other novels by Dick: "We may feel that both novels [*Martian Time-Slip* and *Clans of the Alphane Moon*] are intensely concerned with sterility, as Dick's works invariably are, and that this sterility has an ontological dimension" (Palmer 222–223). That sterility — and it would be a revealing exercise to check how often Philip K. Dick uses words like "barren" — lies not in the planet but in the inability — we could say *acquired inability*, as the causes are above the individual per se — to establish empathic relationships with his fellows, human or otherwise.

Kim Stanley Robinson, climbing a few steps higher than his own previously quoted description, elaborates also on that dominance of economy over the individual: "naked capitalism" (Robinson 58) turns the planet into "the ultimate consumer society" (Robinson 56), a "nightmare reality" (Robinson 58) where only the most discerning — in which case, one has to be either schizophrenic or autistic — see nothing but "gubble." One can, of course, try to profit from the situation and succeed for a while, like Arnie Kott, Leo Bulero, or even Otto Zitte, or commit suicide when failure comes, like Norbert Steiner. Ultimately, however, that failure, just like entropy, will catch up. The only ones who seem to be able to escape are those who, from the beginning, are estranged from Earth's economic and political value system — that is, only the native Martians. They may still have to endure their condition as "colonized" (just like the "bleekmen" in *Martian Time-Slip*) or be caught and instrumentalized in the unstoppable wave of commodification (arguably like the "papoola" in *The Simulacra*), but when all things pass and corporations like AM-Web are nothing but a pile of old buildings, Nature will take over. Nothing short of an apocalypse — even a secular one — is potent enough to overcome the slide from unrestrained capitalism to a tomb world.

Specters of Mars

We must nevertheless note that in *The Simulacra* and also in *Do Androids Dream of Electric Sheep?* Mars is not a setting for the novel's events, but rather a background that either fades out to give way to a more Earthly narrative (*Do Androids Dream...?*) or slowly fades in so that, in the end, unlike those "Martian novels" where the planet is merely an extension of Earth's dystopia,[3] a few characters may embrace it as a refuge (*The Simulacra*). Robinson does not seem to agree with that (maybe naïve) view: for him, Philip K. Dick's Mars seems so coherent in other novels[4] that such an anomaly has to be taken care of by a more attentive and informed reader:

> Three of the most sympathetic characters escape at the end of the novel [*The Simulacra*] in a spaceship—but note where they are headed.... This is the only hopeful moment..., but because they are escaping to Mars, we must pause to wonder if Dick means this ironically ... we [are] left with what we know of Mars from Dick's other works [Robinson 71–72].

We can, however, suspend Robinson's claim and start by assuming that each novel is singular, even if a consistent picture subsequently emerges. In this particular case, one of the unique features of *The Simulacra* may be precisely the fact that Mars is still a place where hope can be sought, a place to establish a resistance movement or, at least, to flee temporarily from the crumbling dystopian regime and the emergent triumph of the "chuppers." In the end of *The Simulacra*, we can only speculate on what would be the life of those characters who escape to Mars—Al Miller and Ian Duncan, along with Loony Luke—but a few other novels (even if not "Martian") may give us some hints.

The "settlers" of *Solar Lottery* end up stranded in the mythical tenth planet, where they discover that what remains of Prester John is nothing but a recording, along with their own religious (and therefore futile) wish fulfillment; grim survival is the only thing they can expect. For the proto–Venusians of *The World Jones Made* there is a "new hope" for a genetically modified humankind, but these will be, after all, estranged from our planet as much as we are from their future home. Analogous to *Solar Lottery* but unlike Robert Heinlein's *Stranger in a Strange Land*, there is no expectation of a cultural clash to modulate that estrangement into something new. A better comparison point may be the midpart of *The Man Who Japed*, when the protagonist Allen Purcell is sent against his will (and against his knowledge) to a planet in the Vega system known as "Other World." Supposedly in a world where his deepest desires are fulfilled but really in a "Mental Health Resort," Purcell could have preferred this new "perfect" life to the one where he unconsciously desecrated the statue of the regime's father figure, Major Streiter. After becoming aware of the ersatz quality of that world, there is no vacillation: return to Earth becomes an imperative to be carried out by all means, legal or otherwise. Despite what may await him, his words are resolute: "It's good to be back" (*MWJ* 87).

Just as in the folk *récits de quête* examined by Propp, and particularly in the further systematization by Greimas, this "Other World" is a paratopic space[5] where the main character acquires the competences that will be needed in his final quest. In this specific case of *The Man Who Japed*, competence equals maturity and knowledge. His early "japing" was a repressed desire; from then on, his acts become deliberate, culminating in Dick's tribute to Jonathan Swift's *A Modest Proposal*, near the end of the narrative. In other occasions—hence the comparison between this earlier novel and *The Simulacra*—awareness of a dystopian regime already exists, but the characters have no means to overcome it. Exile, short or of undetermined length, may be the better choice. Nevertheless, a return to Earth is always in their minds, as long as it remains suitable as a planet to take care of.

"Still a planet to take care of and supervise"

Every now and then, Philip K. Dick's ecological concerns have been noticed, namely in shorter titles such as "Autofac" (cf. his own introduction to the 1977 collection *The Golden Man*, evoking Thomas Disch) or such novels as *The Game-Players of Titan*, not to mention essays like "The Android and the Human." Those concerns make sense, however, only when framed in the wider context of economy and politics in his work. What is at stake in "Autofac," for example, is not the environmental damage of the unstopping automated factories by itself, rather the fact that an untamed dominance of corporations leads to untamed production, which in turn leads to an untamable depletion of natural resources.

If a sustainable solution is to be applied to the causes of these typically Dickian predicaments, instead of a mere palliative to the consequences, it becomes clearer why our home planet matters so much more than Mars. Better: why that planet (however it is portrayed) should be understood as a mirror of Earth (as Kim Stanley Robinson argued), but also, although there is nothing playful in colonization or in exile, as a big sandbox. In the Mars of *Martian Time-Slip*, we have — note the uncanny congruence with the "slippage" in the title — simultaneously our past (the frontier land, the depletion of the resources and exploitation of the natives), our present (a commodified society, the dehumanization of individuals, due, for example, to the robot "teaching machines"), and our future (the desolate landscape dreamed by Manfred, even if in the end brighter possibilities emerge).

Other novels are less explicit in the narrative function of the Red Planet, but by now we may dare a hypothesis. In *Do Androids Dream of Electric Sheep?*, Earth became almost inhospitable for humans, who have migrated to Mars. It is up to our "mirror images," the androids — at least in the way Dick portrays them[6] — to be bold enough to make the opposite movement, as if feeling more nostalgic than us for a homeland that they barely knew. After having experienced the "simulacrum-planet," they are looking after the real thing. Humankind, in its turn, is divided between those who have moved to Mars and those who are forced to stay; part of the tragedy lies in the impossibility of a (political) cooperation with androids, which remain on the borderline between being a workforce devoid of citizenship and nothing more than criminals.[7] A large-scale solution, apart from the harmony within himself that Rick Deckard finds at the end of the novel, would require that settlement of the differences between humans and androids. Mars thus represents a failed solution, one where inequality was the rule and androids were the exploited; on Earth remain the humans that could — but still don't want to — find in androids equals among the wretched.

The reader does not know, in *The Simulacra*, what kind of life is there on Mars for the characters who flee from the crumbling totalitarianism at the disenchanted close of the novel. There is, however, some hope in that disenchantment: The exile seems long but temporary, and those who will take hold of the Earth, the "chuppers," are at least in tune with Nature, even if not with our social norms. That ability to be in tune — first of all with Nature, but, by extension, also with a community of fellows — is, remarkably, a feature often associated, in science fiction, with alien planets and their inhabitants, and Mars is no exception. We simply need to evoke two well-known cases, C.S. Lewis' *Beyond the Silent Planet* and Heinlein's *Stranger in a Strange Land*, to assert that generic convention. In both novels, the Martians stand for that "ecologic" affinity with Nature — or even some kind of Divinity, in the case of Lewis — that humans seem to have lost.

Although some singular features in each of those novels are less relevant to our argument, it cannot be left unnoticed that, in *Stranger in a Strange Land*, Valentine Michael Smith, biologically a human but a Martian by nurture, is able to perform a Hegelian synthesis, overcoming the antithetical features of both cultures, and showing humans new paths, even if his proposals are blasphemous from a purely "humanist" perspective. In *Beyond the Silent Planet*, that all-too-human "fall from Grace" seems however to be irrevocable, apart from a few chosen ones like Ransom who are qualified for remission.[8] In spite of Dick's religious faith, that is a solution we do not find, not even in his later novels.[9] We can, nevertheless, admit some similarities with that rhetoric device: The future Mars from where *Martian Time-Slip*'s Manfred sends his message of hope could also be Earth's future, the future that lies beyond the open ending of *The Simulacra*. Unlike Heinlein, however, adaptation and not synthesis is the rule: Manfred Steiner becomes a Martian, if only because he is already alienated from Earth's culture; the characters from *The Simulacra*, if and when they return, will be ready to start over, in a more humane — and perhaps humanist — way.

Maybe we can, after all, reestablish the coherence across Philip K. Dick's work, but at a higher level than the one proposed by Robinson. Preceding the "Mundane SF" manifesto, Dick's political approach to fiction denotes — perhaps with the exception of some more formulaic short stories and novelettes of his formative period of the 50s, or *The World Jones Made*, mentioned above — a need to solve Earth's problems on our native planet. A temporary "retreat to the forest" — I am thinking here of Ernst Jünger's essay *Der Waldgang*, but also, as previously brought up, of some concepts of Greimas' narrative semiotics — can sometimes be the most appropriate strategy, and Mars can be that forest where we remain hidden, waiting for the appropriate time to return to the *Heimat*-planet that is and always will be Earth.[10] Even in *Do Androids Dream...?*, as suggested above, the fulcral position of Earth is testified, in one of those role reversals that have given the novel its status as a classic, not by humans but by the androids. The return to our (and *their*) planet, regardless of more instrumental reasons, also has the symbolic meaning of homecoming. Maybe not a prodigal son's return, but a return nevertheless.

This means that there may be a fourth option besides the ones enumerated by Hazel Pierce in "Philip K. Dick's Political Dreams":

> How can one escape [in *The Simulacra*] this mad absurdist world? It might be possible to migrate to a virgin environment and attempt to reconstruct from the bottom up the very same society you escaped, improving it through the hindsight of prior experience. Or one might retreat into social invisibility, living a life of psychological isolation and tainted freedom in the midst of the crowd. A third alternative is the complete destruction of civilization so that a more primitive social group may have the chance to set the stage with new hopes and new directions [Pierce 128].

If — with the much-needed prudence — other novels can be used as hints about which of those alternatives is the most "Dickian,"[11] migration evokes the Prestonites from *Solar Lottery*, and the building of a new civilization (after the abandonment or destruction of the old) shares some traits with the end of *The World Jones Made*. Isolation (or an inner retreat), while being particularly common when dystopias appear with their darkest tones — *The Man in the High Castle* being their epitome — is something more legitimate if assigned to single characters than to the "sense of an ending" of the novel as a whole. Dick may cover all those alternatives, but underlying all of them is that other belief that, just like in Leo Bulero's memo that opens *The Three Stigmata...*, no matter how bad things go, "we're not doing too bad ... even in this lousy situation we're faced with we can make it" (*3SPE* 2). Earth may also be a "planet for transients," but it is our own, something that will always be lacking in Mars.

Notes

1. Although discussed in the same chapter of Kim Stanley Robinson's thesis for chronological reasons (cf. Robinson 51–64), the epithet fails in the case of *The Game-Players of Titan*.

2. That very same 1950s normality is criticized in *Time out of Joint*, as Fredric Jameson noted in his *Postmodernism, or The Cultural Logic of Late Capitalism* (279–280).

3. That very same dystopia that Robinson identifies as "the most common element in all of Dick's work" (27).

4. To the point of *The Three Stigmata...* being defined as the same planet as *Martian Time-Slip* after its economical expansion (cf. Robinson 60).

5. Cf. the entries "Paratopic," "Topic Space" and "Utopic Space" in Greimas and Courtés's *Semiotics and Language: An Analytical Dictionary*. The initial predicament for the narrative occurs in the "topic space"; "paratopic space" is where the main character acquires the competences needed to overcome that predicament, and the performance that allows a new steady state of affairs usually takes place in the "utopic space." Our claim is that — *mutatis mutandis*, as Dick's novels should not be taken as popular "marvelous" narratives — planets such as Mars typically embody that "paratopic space." One of the best illustrations can, however, be found not in Mars but in Whale's Mouth, in the novel *The Unteleported Man*.

6. Instead of retelling here Dick's conception of what it means to be human, it is enough to evoke his two essays "The Android and the Human" (written in 1972) and "Man, Android and Machine" (1976).

7. Cf. Giorgio Agamben's *Homo Sacer: Sovereign Power and Bare Life*, on the "naked life" condition of the accursed, the political refugees, etc., following Carl Schmidt, Hannah Arendt and Michel Foucault. All the worse for Dick's androids, in spite of his redefinition of "humanity": In the novel, their right to political life [*bios*] is denied because they lack a biological life [*zoe*] to start with. As an illustration of this ideological entanglement, the Voigt-Kampff test — as Jill Galvan argues in "Entering the Posthuman Collective in Philip K. Dick's *Do Androids Dream*

of Electric Sheep?"—is nothing more than a "linguistic apparatus [that ...] assures the android's condemnation" (421) even before the actual test is performed.

8. In C.S. Lewis (almost in a diametrical opposition to Heinlein's playful subversion of contemporary conventions), humanism equals remission, a presupposition that is reinforced by Ransom's skill as a linguist, opposed to Weston's and Devine's Faustian preferences for an unethical approach to science and technology. But, as Robert Scholes has noticed in "Boiling Roses: Thoughts on Science Fantasy," it is from the beginning a humanism modulated (or perhaps contaminated) by "not just Christianity, but a very Catholic version of it" (18).

9. Cf. his capacity for self-irony, for example in the hilarious passage in *VALIS* where Horselover Fat talks to the fundamentalist therapist Maurice (*VALIS* 96–97).

10. Unless we no longer feel it like ours, as the "lunatics" in the novel *Time Out of Joint* or, to a certain extent, the Venusians in the novelette "War Veteran." Both were written in the fifties, though, a fact that weakens their strength as a counter-example.

11. Maybe all of them are. The fact that all (re)appear in *The Simulacra* also reinforces Kim Stanley Robinson's claim on the novel being one illustration of his "extravagant inclusion" of too many elements (cf. Robinson 72).

Works Cited

Agamben, Giorgio. *Homo Sacer: Sovereign Power and Bare Life*. Stanford, CA: Stanford University Press, 1998.
Aldiss, Brian. "Dick's Maledictory Web: About and Around *Martian Time-Slip*." *Science-Fiction Studies* 5 (vol. 2, pt. 1). March 1975.
Dick, Philip K. "The Android and the Human." *The Shifting Realities of Philip K. Dick: Selected Literary and Philosophical Writings*. Ed. Lawrence Sutin. New York: Vintage Books, 1995.
_____. *Do Androids Dream of Electric Sheep?* In *Philip K. Dick: Five Great Novels*. London: Gollancz, 2004.
_____. "Man, Android and Machine." *The Shifting Realities of Philip K. Dick: Selected Literary and Philosophical Writings*. Ed. Lawrence Sutin. New York: Vintage Books, 1995.
_____. *The Man Who Japed*. In *Three Early Novels: The Man Who Japed, Dr. Futurity, Vulcan's Hammer*. London: Millennium/Gollancz, 2000.
_____. *Martian Time-Slip*. In *Philip K. Dick: Five Great Novels*. London: Gollancz, 2004.
_____. *The Three Stigmata of Palmer Eldritch*. In *Philip K. Dick: Five Great Novels*. London: Gollancz, 2004.
_____. *VALIS*. London: Gollancz, 2003.
_____. *We Can Build You*. London: Voyager/Harper Collins, 1997.
_____. *The World Jones Made*. New York: Vintage Books, 1993.
Galvan, Jill. "Entering the Posthuman Collective in Philip K. Dick's *Do Androids Dream of Electric Sheep?*" *Science Fiction Studies* 73 (vol. 24, pt. 3). November 1997.
Greimas, Algirdas, and Joseph Courtés. *Semiotics and Language: An Analytical Dictionary*. Bloomington: Indiana University Press, 1982.
Heinlein, Robert A. *Stranger in a Strange Land (New Edition)*. New York: Ace, 1987.
Jameson, Fredric. *Postmodernism, or The Cultural Logic of Late Capitalism*. Durham, NC: Duke University Press, 1991.
Lewis, C.S. *Out of the Silent Planet*. New York: Scribner, 2003.
Pagetti, Carlo. "Dick and Meta-SF." *Science-Fiction Studies* 5 (vol. 2, pt. 1). March 1975. Trans. by Angela Minchella and Darko Suvin.
Palmer, Christopher. "Critique and Fantasy in Two Novels by Philip K. Dick." *Extrapolation* 32.3, Autumn 1991.

Pierce, Hazel. "Philip K. Dick's Political Dreams." *Philip K. Dick*. Eds. Martin Harry Greenberg and Joseph D. Olander. New York: Taplinger, 1983.

Robinson, Kim Stanley. *The Novels of Philip K. Dick*. Ann Arbor, MI: UMI Research Press, 1984.

Scholes, Robert. "Boiling Roses: Thoughts on Science Fantasy." *Intersections: Fantasy and Science Fiction*. Eds. George Slusser and Eric S. Rabkin. Carbondale: Southern Illinois University Press, 1987.

Sutin, Lawrence, ed. *The Shifting Realities of Philip K. Dick: Selected Literary and Philosophical Writings*. New York: Vintage, 1995.

KIM STANLEY ROBINSON
From *Icehenge* to *Blue Mars*
Christopher Palmer

Kim Stanley Robinson has remarked that it took a while for SF writers to respond to the 1976–80 revelations about Mars. Clearly, he could be confident that he, at least, had responded by 1993–96, when there appeared his big Mars trilogy, full of work, journeying, geology, and politics. But things were different earlier, in 1984, when Robinson's first Mars novel, *Icehenge*, was published. In describing *Icehenge* below, I keep in mind the later Robinson, rolling triumphantly from the Mars trilogy to the recent Climate trilogy. In this later Robinson the energy and zest, which carries both writer and readers through some thousands of pages, comes from an alliance between the characters' activity of knowing and talking and doing and hiking, and the author's creativity in imagining all this. It's a classic SF effect. World-embracing and, latterly, fearful problems figure as challenges and opportunities. The mill seems big enough for any grist.

Icehenge is in contrast with this. In *Icehenge* knowledge is uncertain, hard to fix; the depths of personality which should lend energy to the seeker and doer tend to neurosis. Yet in fact the matter is more complex. The darknesses and gaps to be found in *Icehenge* persist in the Mars trilogy; at times they haunt the later work, though at times it finds ways of confronting them. The final part of this essay will touch on this complexity. The crucial feature is a clash and also a dialectic of transparency and non-transparency.

In *Icehenge* and the first "Green Mars"[1] the Mars project, to which the Mars trilogy will devote such enormous detailed energy later, has already failed, politically and socially, with the dubious exception of the longevity treatments — dubious because the gift of very long life, already present in the earlier fictions, mainly figures as opening the way to disillusionment, and to amnesia. "What remains of the distant past is jammed under the weight of subsequence, so that recollection is stressed, then disabled" (*Icehenge*, 122: an interesting geological image, because geology is usually transparent in Robinson. Geology is how a landscape exhibits its history and, so to speak, its experiences).

Mars is ruled by a tyranny which controls information and rigs history. Settlement must have been an achievement, but the history of that achievement has been lost and perverted.[2] So, perhaps, progressive movement has to be outwards, elsewhere to where a better future might be made. The Davydov expedition, launched in mutiny against the ruling Mars Committee, did move outward, leaving behind it Icehenge, a monument on Pluto with a motto in Sanskrit meaning "to push further out," and the date of departure from our system, 2248. The rest of the novel, however, is devoted to revelations and complications which leave the ideal of the expedition in disarray, substitute a circular or recursive pattern with each successive protagonist recycling obsessions with earlier ones, and suggest layers of fakes around the Icehenge monument.

The story is told by three successive narrators, and in the first person (unusual in Robinson), and their subjectivity is important. There are mirror moments for each—moments when a look in the mirror (or into a transparent wall) shows an alien self.[3] *Icehenge* is a rather unsocial book; solidarity and collective cooperation largely disappears from the story with the departure of Davydov and his companions. Emma, Hjalmar and Edmund are all, in their ways, loners. Emma is an expert in making systems as far as possible closed ones; that's how she got involved with the Davydov expedition, to help them seal their ship against energy loss, and the three main characters tend to be closed on themselves, living in a circle of haunting lover-figures or father figures. The lovers are yearned for, missed, possibly fantasized; the fathers are the subject of an antagonism by means of which the individual in question structures his life (Hjalmar against the patron and lover he calls the Shrike; Edmund against Hjalmar, who is his grandfather). The monument itself, a circle of ice pillars, comes to seem closed, at least in the sense of being enigmatic. It proliferates theories, rather than proclaiming an idealistic message. The force of its first message — the first *interpretation* of its message — is lost.

Emma leaves behind a narrative of her role in the departure of the Davydov expedition and then in the failed revolt on Mars against the committee. She perhaps disappears into the Martian wilderness, perhaps to live on. Hjalmar, an archaeologist, embittered and neurotic but stubborn, seems years later to have recovered the truth of the revolt, and Emma's escape, and the truth of Icehenge and the Davydov expedition. In this middle section the arc of the story seems to reach up, though not very obviously, as Hjalmar is an unattractive hero, even to himself. Hjalmar is haunted by the idea and perhaps even the presence of Emma. In a crucial episode (148–163) he ventures into the wilderness. Seemingly led on by Emma, he is guided to a kind of pavilion, refuge of the escaping rebels, whereupon Emma and the pavilion disappear overnight. The experience cannot be scholarly evidence for anything to do with Emma and the rebels, but it forms part of an ordeal that appears to redeem Hjalmar. (Robinson remarked that this section, written last, was the most satisfying for him. The episode of Hjalmar's ordeal in the wilderness might be one reason for this: It's very Robinsonian, and anticipates the repeated, alert-immersed trekking of the trilogy.)

After this, in part three, the story heads downwards. We are in the hands of Hjalmar's grandson Edmund, who is cheerfully flippant and apolitical rather than bitter and neurotic. His attitudes suggest that the denizens of this further future have lost the sense of political possibility that still survived in Hjalmar's time. He sets himself to destroy Hjalmar's explanation of Icehenge, and in effect appropriates his grandfather's obsession with Emma, whom Edmund now meets in dreams, and whom he also transmogrifies into the reclusive millionaire Caroline Holmes. According to Edmund, there was no Davydov expedition. It was Caroline who erected Icehenge, and the date on it commemorates her year of birth, not the year Davydov and his companions left our system; the whole thing is a kind of hoax. In fact, Caroline is very possibly Emma herself (if the pronoun makes any sense by this stage), living on, though Edmund provides this reincarnated Emma with even less motive for the fake than he found for Caroline before she became Emma in his theorizing. There are some holes in this, but the combination of a kind of appropriation, and a layering of theories and texts so that the truth is left uncertain, is effective. And then Robinson inserts a discovery, a chamber under Icehenge that could have been erected by Holmes but not by Davydov, to tilt conclusions towards Edmund's theory.

The result is a rather grim account of the process and progress of knowledge. Even after Hjalmar's apparent achievements, knowledge is vulnerable to dispute, appropriation, fakery, individual obsession. History can't be conclusively recovered so that it might serve as a stimulus to future action; the past can't be remembered (because of the amnesia of longevity) but it haunts, or lives on, in Emma/Caroline as they exist for Hjalmar and Edmund, in unreliable yearnings and elusive apparitions, and in Edmund's Oedipal drives. Even Edmund turns in on himself in the sense that his remarks about hoaxers tend to apply equally to himself, lessening the difference between the hoaxer and the supposed uncoverer of the hoax. The would-be uncoverer of the hoax seems to have transferred to the alleged hoaxer his own desire to own the story.

Now, this summary account of the novel probably exaggerates its engagements with skepticism and indeterminacy. It overlooks the engagement with the landforms, the geology of Mars, in the ascent of Olympus Mons in "Green Mars" and in Hjalmar's ordeal in the wilderness in *Icehenge*: journeying on Mars as the life of full practical consciousness, and the landscape of Mars as transparent, as the visible history of the planet, to which human awareness can then serve as memory, even in humans whose own memory is fading. These are motifs to be developed in the Mars trilogy. Nonetheless, the overall narrative of *Icehenge* is shaped by the indeterminacy of memory and story, and that is interesting in this author. When we get to the end of the novel we can still go back from Icehenge itself, so thoroughly problematized, to the pavilion in the wilderness that Hjalmar encountered (156–157), or to the ruined lamasery that Roger and his companions find at the top of Olympus Mons at the end of the first "Green Mars" (111): other ruins, other messages. But it's interesting to take stock of this postmodern and disturbed moment early in Robinson's career, especially as it persists, for instance in *A Short Sharp Shock* (1990) in which a person, unnamed, not individualized, amnesiac, travels a planet on a central spine of land which circles the planet, only to arrive at his starting point and begin his journey again by again plunging through a mirror. This novella is so ferociously nonlinear that it seems as if it was Robinson's attempt to get circularity out of his system for good and all.

In much of his fiction between *Icehenge* and *Red Mars*, Robinson can be seen as working some of these tropes out of his system, or simply as working on them. There is an engagement with the past as it might survive into the present in monuments and memorials, in stories into the '90s, and indeed in episodes written later, catching the dark or challenging presence of the past in the present, as we see in several meditations on the Vietnam War memorial and several incidents which cast the Lincoln Memorial in a sobering or an uncanny light.[4] There's an analogy between the darkness of the historical past and that of the individual past, as figured in the amnesia that afflicts the long-lived in *Icehenge*; later stories do start to disentangle the two, whereas the personal absorbed the historical, the monument was appropriated, in *Icehenge*. Thus "A History of the Twentieth Century, with Illustrations" (in *Remaking History*, 1991), in which the personal message of the Vikings who once stayed in the stone tower comes through clearly to the depressed protagonist: "Ingrid is the most beautiful girl in the world," say the runes (85). Similarly, the paired stories "The Lucky Strike" and "A Sensitive Dependence on Initial Conditions" meditate on how history might be open, or made open; and "The Translator" makes wonderful comic play with cunningly distorted information, the truth better kept from those who might do harm with it — questions of secrecy that receive a grim treatment in the later Climate trilogy.[5] Secrecy and openness, haunting mystery, sheer blankness, the productivity of lucid description and understanding; the past as oblivion, a gap in memory like a trauma such as the main character has suffered in "Ridge Running," history as able to be seized, its potentials released, by individual or collective action, the future as a set of possibilities to be worked on: Robinson's fiction continues to worry at these factors and aspects.

It was emphasized that the characters in *Icehenge* were troubled loners, and suggested that their isolation allowed their searches for the truth of the past to become raveled by their needs and fantasies. It looks as if Robinson turns decisively away from this kind of thing in his later, grandly planned and world-shaping novels. He is intensely interested in the social life of science and even of bureaucracy, the free exchange of thoughts and information, "so public, so explicit" (*Blue Mars*, 677). His characters are commonly active, sociable, engaged. It can seem that the full life of any individual is in this activity and engagement. Yet the characters' social busyness often topples into doubt and depression: where in all this activity and engagement is the self? You are fully yourself because you are fully engaged, but the self is also suspended in the engagement. There are some episodes of hyperactivity, when extroversion starts to seem unhinged, with John Boone and Frank Chandler (*Red Mars*— the latter "manic with overwork," 250), and with Frank Vanderwal (*Fifty Degrees Below*). Humans do move onwards, as Robinson's oeuvre has, attentive to nature's disclosures of itself, absorbed in the task; there's a fullness of the orientation to the future that is a part of SF (and of being modern and maybe of being American); but the past still haunts the fictions and the characters, and absorption in the task can be a threat to the subject.

The situation in Robinson's oeuvre (so far) as a whole can be seen in terms of an underlying contrast between the transparent and the nontransparent. In general, but especially in his later works, Robinson loves talk and sociability; he values the "dialogic," as Carol Franko has said (1997), "the age-old excitement of honest debate" (*Red Mars*,

316); his characters have often a kind of innocent, buoyant candor about them (perhaps this is why they are so often named Frank)[6]; he dwells in detailed description; he's a very *informative* writer; and science has to be (at least for other scientists) transparent so its discoveries can be tested and repeated. And transparency is also a democratic ideal, manifested in the politics of the Mars trilogy.

What we want of human beings, in most of Robinson's fiction, is that they exchange with us and disclose to us; but the novels are haunted by mysterious women who tantalize by withholding or disappearing: the main character's mysterious anima in *A Short Sharp Shock*, Emma and Caroline in *Icehenge*, Hiroko in the Mars trilogy,[7] Caroline in the Climate trilogy. This later Caroline is forced to be elusive, by her husband, who is evil in the way he steals yet withholds information: bad nontransparency. Evil is rare in Robinson, a novelist of redemptions and recoveries (for instance that of Frank Chalmers, in the Mars trilogy, a murderer redeemed). And we scarcely get to look Caroline's husband in the face. We come to some sort of crux or even aporia of subjectivity here, as this species of woman figures as completely other (you can't know her or catch her) but may well be the creation or illusion of the man who yearns for her. The Caroline of *Fifty Degrees* and *Sixty Days* shifts between being an elusive creature of desire and yearning, and a warm flesh-and-blood person almost desperately held by Frank Vanderwal for an hour or so, but still, even then, elusive in her plans and purposes.

We have to think about art in this broad contrast of transparent and nontransparent in Robinson's fiction. Art is different from science as regards transparency. Art resists paraphrase; there must be a residue, something not quite paraphraseable, and we suspect that the art of art lies precisely there. It's not replicable. It's nontransparent. *Icehenge* acknowledges this; the eponymous monument (a work of art, after all) remains a mystery; or perhaps it becomes a different thing, a puzzle, and the mystery is elsewhere in the novel, in the lost, yearned for, hallucinated Emma Weill.

The later fiction repeatedly finds ways of acknowledging this, though not, in my opinion, in the works of art actually imagined, for instance, in the Mars trilogy, which tend to be benign in a communal way. Much more important is the continuing interest in the sublime, and in the epiphany or (in the Mars books) areophany. This is the lone witness, struck and transfixed by some moment supercharged with meaning that can't be easily exhausted and almost overwhelms: "pure sublimity," "nostalgia to the nth power," "heartbreaking, or heartfilling" (*Blue Mars*, 723). Sax is lost in a snowstorm and thinks that the vanished Hiroko has guided him to safety with a hand on his wrist; Ann is exploring the newly icebound shores when she is pursued by a bear, like a polar bear only faster, which some free-enterprising terraformer has let loose in the land, and which can symbolize for her the evils of the process; Michel, characteristically, has his areophany on Earth, when he is visiting Switzerland and realizes from the vastness of the Alps that "Earth is a greater Mars." You need mind to bring about the sublime, as Caroline Holmes says in *Icehenge* (225); it requires conscious human presence[8]; but at moment of the sublime, for instance when experiencing epiphany or areophany, the mind stops, and the experience can't be repeated or explained. You feel awe, or you feel the mysterious for-itselfness of whatever it is that has epiphanically appeared.

In one way, the activity of the Mars trilogy is the same as that of the Martians in it: a struggle towards progress by work and talk, whereby they *become* Martians; in another way, it is to issue in the moments of areophany at which talk falls silent (except, of course, that these moments are described in the text) and the moment's progressive potential is irrelevant. The universe needs humans, who bring consciousness to it; but then almost at once they change it, act on it, and it begins to change: "And all these things went away" is the refrain of the history of Orange County in the interchapters in *The Gold Coast* (1988), and it's a constant in Robinson's writing: "Nothing lasts, not even stone, not even happiness" (*Red Mars*, 109). Everything has to be registered, described (and understood), and this drives both the writing and the scientists and observers who populate the fiction, because it begins with naming ("naming was the power that made every human a scientist of sorts," *Red Mars*, 53); but nothing remains once (modern) humans arrive. And, further, the dark shadow of epiphany, especially in the Mars trilogy, is amnesia, the lapse of memory in which nullity enters, the "blankout" (*Blue Mars*, 713). There's a pattern here which is powerfully abyssal: that is, one kind of limit to words and thoughts, amnesia or nullity, appears beneath another, the sublime or areophanic.

What happens towards the end of *Blue Mars* is that the threat from the loss of memory in the long-lived, a presence in the Mars books since *Icehenge*, becomes stark:

> Mind was one's body's life. Memory was mind. And so, by a simple transitive equation, memory equalled life. So that, with memory gone, life was gone [*Blue Mars*, 687].

This is accompanied by a renewed feeling for transience:

> ... everything was passing—everything they did was the last time they would ever do it [*Blue Mars*, 622].

It is as if certain threats and pressures come to the fore when the great work has largely been achieved. Thus starkly challenged, however, the energies of Science are summoned up, for Sax in particular, and the trope of attentive journeying on Mars now turns inwards, to the interior of the brain, and with some success too.

This essay, then, defines a few of the things that trouble the living stream of Robinson's positive, problem-solving, wondering fiction, using *Icehenge*, where these things tend to dominate, and suggesting that a large part of the interest of his later texts is in how they are challenged by and deal with these other elements. Their technological and political energy and extroversion is complicated and deepened by this sense of the unstable, withheld, and resistant to description; in that case the narratives work dialectically, deepening and reflecting as they narrate expansion and achievement.

Notes

1. That is, the novella of 1985 rather than the novel of 1994 which is the second in the Mars trilogy.
2. Clearly, the Mars trilogy will tell the story of this achievement in detail; but it begins with a murder, that of John Boone, instigated by his dark brother Frank Chalmers, an echo of the murder of Remus by Romulus at the founding of Rome (*Red Mars*, 4–22).

3. See *Icehenge* 25 (Emma), 112 (Hjalmar), and 223 (Edmund); the image also occurs in one of Robinson's '90s interviews about authorship, expressing the idea of an other self that dictates: "The Profession of Science Fiction, 34: Me in a Mirror," *Foundation*, 58–63.

4. Vietnam Memorial: "The Memorial," *In the Field of Fire*, edited by Jeanne Van Buren Dann and Jack Dann, 1987, and in "A History of the Twentieth Century, with Illustrations"; Lincoln: "A History..." again, and in *Fifty Degrees Below*.

5. "The Lucky Strike" in *The Planet on the Table*, 1986; "A Sensitive Dependence on Initial Conditions" and "The Translator" in *Remaking History*, 1991.

6. "I have always used the name Frank for the characters I think of as most two-faced. This is the kind of primitive joking I indulge in when naming characters and I like repeating names and using heavy-handed symbolism there; why not, and where else can you get away with it so well" (Robinson in an e-mail, 2008).

7. Another who sets out for the stars, according to rumor at least (*Blue Mars*, 654).

8. We need to make Earth into World, as is said in "Green Mars."

Works Cited

Franko, Carol. "Utopian Destiny in *Red Mars*." *Extrapolation* 38, 1, Spring 1997.
Robinson, Kim Stanley. *Blue Mars*. London: HarperCollins, 1996.
_____. *Fifty Degrees Below*. London: HarperCollins, 2005.
_____. *Forty Signs of Rain*. London: HarperCollins, 2004.
_____. *The Gold Coast*. New York: Tor, 1988.
_____. "Green Mars." New York: Tor, 1985.
_____. *Green Mars*. New York: Bantam, 1994.
_____. "A History of the Twentieth Century, with Illustrations." *Remaking History*. New York: Tor, 1991.
_____. *Icehenge*. 1984. Reprint, London: HarperCollins, 1997.
_____. "The Lucky Strike." *The Planet on the Table*. London: Futura, 1986.
_____. "The Memorial." *In the Field of Fire*, edited by Jeanne Van Buren Dann and Jack Dann, New York: Tor, 1987.
_____. "The Profession of Science Fiction, 34: Me in a Mirror." *Foundation*, Volume 38. Winter, 1986–87.
_____. *Red Mars*. New York: Bantam, 1993.
_____. "Ridge Running." *The Planet on the Table*, London: Futura, 1986.
_____. "A Sensitive Dependence on Initial Conditions." *Remaking History*. New York: Tor, 1991.
_____. *A Short, Sharp Shock*. Shingleton, CA: Mark V. Ziesing, 1990.
_____. *Sixty Days and Counting*. New York: Bantam, 2007.
_____. "The Translator." *Remaking History*, New York: Tor, 1991.

Martian Musings and the Miraculous Conjunction

Kim Stanley Robinson

The fact that Percival Lowell really believed in his vision of canals on Mars — and then also the resulting scenario of the dying civilization that was trying to move water from the poles to the equator — means that Lowell wasn't really a hoaxer, even if his vision had the same effect as a hoax on the public. It's just that Lowell also was deluded and seduced by his own story. Astronomers like Hale and Barnard, who were editors of the astronomy journals of the time, would not publish Percival Lowell's work. *They* knew a science fiction writer when they saw one. In fact Lowell was restricted to self-publishing his work, and he was one of that subset of science fiction writers who believed in the reality of his own creation.

But the impact of his announcement to the world of canals on Mars was extraordinary. A hundred and ten years ago it was gigantic news, which galvanized the writers of the world. At that time there wasn't a science fiction establishment like there is now, but nevertheless there were writers perfectly happy to write science fiction when an idea as good as this one came along. For me the most important one is Aleksander Bogdanov's *Red Star*. Published in Russia in 1905, it had an enormous impact, sold several hundred thousand copies, and inspired a whole generation of Russian revolutionaries, some of whom also became scientists and founded the Soviet space program. Then the same thing happened in Germany with Kurd Lasswitz's *Two Planets*. There was an entire book club devoted to that one book, creating a movement that led to the German Rocket Society, to Willy Ley and Wernher von Braun and others. A very strong argument can be made that we would not have gotten to the Moon by 1969 if it weren't for Percival Lowell's fantasia about Mars, because you wouldn't have had the German Rocket Society as a result of *Two Planets*. So, it being the case that we've never been anywhere else but the Moon in 1969, one could say that we wouldn't have gotten off this planet if it weren't for a science fiction writer who thought he was a real astronomer.

In Britain, H.G. Wells took Lowell's idea and immediately used it as a way to attack British imperialism. He put the imperial forces at the wrong end of the gun and asked what would it feel like to be a Tasmanian when the British arrived or a Filipino when the Americans arrived, and so he created a great meditation on imperialism in *The War of the Worlds*, using the "cognitive estrangement" technique that Suvin has identified as central to the genre, in this case a kind of reversal.

Each one of these books — the Russian one being a communist utopia on Mars, the German one being a technological utopia on Mars, the British one being an antiimperialist diatribe using these hypothetical thirsty Martians — all demonstrate some serious intellectual predilections. To me it's funny therefore that, in America, ten or fifteen years pass before anyone notices this Mars scenario, and then the story is about a cowboy who can leap three times as far into the air as the bad guys, and then afterwards gets to make love to six-breasted women. This is the American response, and it's perhaps as characteristic of our nation as the other books are of theirs. But Lowell, too, was an American, so maybe we have to hold to the idea that the American vision here is not in Burroughs but in Lowell himself.

What all this created was an atmosphere where ordinary educated citizens of the world, for about thirty years after 1895 or 1900, could hold as a perfectly valid belief the notion that there might be a sentient civilization on the next planet out. This is what the scientific community was telling people, as far as they knew, and so it was not a strange belief to hold. But that belief began to go away in the 1930s, because the astronomers themselves kept saying, "Wait a second, there's no Martian atmosphere that we can find, there's no water in the spectroscopes. The planet looks to be dead as a doornail." Orson Welles's famous radio hoax describing a Martian invasion was probably the final blow to this general perception that the Martians might be up there.

For fiction writers this was a bad problem, and resulted in the so-called dry period in Martian literature. The great book of that period is Ray Bradbury's *Martian Chronicles*. What he did was so crucial and fundamental that his book will always exist as one of the great Martian novels. The insight of that book is threefold; just as Asimov's three laws of robotics are famous, there could easily be Bradbury's three laws of Mars, derivable out of the *Martian Chronicles*. The first law is that the ghosts of our Martian stories, whether true or false, are going to be there when we arrive, and they're going to have a huge impact on our minds. The people that go there are therefore going to be permanently haunted by the Martian stories that existed before.

The second law has to do with his Johnny Appleseed story, which is a terraforming story ahead of the fact. It's one of the really fast terraformings, like the one we see at the end of the movie *Total Recall*: thirty seconds and you've got an atmosphere, very convenient. In Bradbury the same kind of thing happens with his character Johnny Appleseeding across the Martian surface and leaving greenery behind. Bradbury, however, calls this "a private horticultural war on Mars," and that's an interesting way to describe terraforming, because it immediately implies the violence of the act to the original Mars, and so it implies also that there will be pacifists in that war, there will be a resistance in that war, and there will therefore be Martian "greens" and Martian "reds." All of that implied future history is right there in Bradbury's phrase "the private horticultural war on Mars."

The third law is simply the famous and beautiful finish to the novel: The father taking his kids down to the canal and showing them "we are the Martians." Human beings will be the Martians. That brings in all of the other elements involved: the reflectivity of it, the fact that you have to look into Martian water, that the water will indeed be there on Mars, but that we will be its actual life. That's a very powerful conclusion.

The dry period went on for a long time, however, after the publication of *Martian Chronicles* in 1950. We begin to see in these years the writers' increasingly desperate attempts to have Martians even though there seemed to be little to no water or atmosphere on Mars. Writers *need* aliens, we need characters, we can't just go to a dead rock — or so it was thought. Aliens are deemed necessary for the sake of suspense, for the sake of an ordinary plot, and so we begin to see in the Martian novels of the Fifties things like sentient tumbleweed — that's one that I particularly like — or the telepathic lichen beds that cover the planet. These are not only telepathic lichen beds — and I like to think about lichen being telepathic — but also malevolent. They're *evil* telepathic lichen beds.

As the dry period went on and the writers got more desperate, the focus shifted from the actual dry Mars to a metaphysically dry Mars, Mars as an image of the wasteland. In these later books it was depicted as close to completely dead, so that there might be things like underground rabbit-like creatures, or perhaps little plastic windmill-creatures, but more and more what Mars came to stand for is simply desolation itself. In particular it became a vision of modernity or suburbia's emptiness, in works like Philip K. Dick's *Martian Time-Slip* and *The Three Stigmata of Palmer Eldritch*, and in the great novel *Farewell, Earth's Bliss*, by D.G. Compton. Compton's novel brings up a really interesting problem, because science fiction so seldom does tragedy, so seldom does wasteland without hope; but this is a mode that Compton never hesitated to attack head on, which may explain why he had a relatively brief (however glorious) career in American science fiction.

At the end of the dry period, just when *Mariner* and *Viking* began to change everything, we were given Frederik Pohl's *Man Plus*, where the notion is that if Mars is a dead rock, what we'll have to do is change human beings to make them appropriate to live in that space. Pohl with his usual honesty took that idea to its logical conclusion, which was that human beings would have to become monsters to exist on Mars. It's a solution. It's not a happy solution, and the result is another somewhat tragic novel.

But what else could you do if there wasn't water and life on Mars? It *was* a tragedy, in story terms. So it was a tremendous relief to all when *Mariner* and particularly *Viking* arrived in the Martian system. *Mariner* in its fly-by had the weird bad luck to photograph only parts of the planet that made it look like our moon. It didn't get any of the obvious riverine features, and so it was really only in 1976, with *Viking*, that the big meteor of goodness hit the Martian story situation, and we got the planet handed to us as if on a platter. No longer was it a little orange dot on which we could project our national fantasies or our personal fantasies. It was another planet, out there visible to the naked eye so we knew it was real, and also now revealed in tremendous detail, with all its stupendous features, many two magnitudes bigger than the biggest equivalent feature on Earth — which was quite a surprise, given that the planet itself was smaller than Earth.

Suddenly there it was, a complete and heroic landscape, and with signs of water too. Almost immediately, by analyzing the *Viking* data, Michael H. Carr of the U.S. Geological Survey wrote the bible on Mars (*The Surface of Mars*, Yale University Press, 1980). For about twelve years it was the basic reference for all Mars students, including in it many surface and satellite photos, with analyses that suggested that if Mars were as smooth as a billiard ball and all its water was up on the surface, there would be an ocean at least 100 meters deep, and perhaps as much as 600 meters deep, covering the entire planet. That was enough water for anything humans might want to do.

So now we had a wet planet. And it's no coincidence that in these very same years, terraforming theory began to be something that the planetary sciences did. Not a coincidence, but a result of the fact that a sudden great candidate for terraforming was shoved right in planetary scientists' faces, and they *had* to think about it. They said, "Geez, there's water there, you could make an atmosphere. There's a lot of land, and there doesn't seem to be any life, so you wouldn't be invading anybody like H.G. Wells's Martians did to us; you'd just have a space that hadn't been gardened yet." Terraforming theory, led principally by Carl Sagan, suddenly became a scientifically viable subject for investigation, and quickly people like Freeman Dyson were suggesting it could be done to Venus too, proposing huge space-cadet methods that would take a hundred thousand years and manipulate enormous physical forces. But with Mars, the first and best candidate, it would take only a little heat, a little bit of pumping, a little bit of human genetic material, and suddenly there would be a living space equal to all the land surface of Earth itself.

That was a very exciting change in the potential of the Martian situation, and I would have to say the science fiction community was not tremendously fast to respond to it. It was a little bit overwhelming. I was there myself, as this is about where I began to enter the Martian picture. When I saw Michael Carr's book it was a mind-boggling experience, because essentially we were being given Martian fiction as it could never have existed before. For me, too, having spent much of my free time in the previous ten years in the Sierra Nevada of California, these Mars photos looked like a place that I recognized. Mars looked like the highest Sierra, the part of it that is above tree line, above about ten thousand feet.

I was wrong in this perception. Mars actually doesn't look anywhere near as much like the Sierra Nevada as it does the Sahara, or the empty quarter of Saudi Arabia — or the Gobi Desert, the Antarctic dry valleys, the Atacama, the Namibian coastline — there are a dozen places on Earth that look a lot more like Mars than the high Sierra does, but the Sierra was the landscape I knew, and that resemblance was what attracted me personally. I thought, you could have a whole planet to backpack on: How good is that? But it isn't really true, and the more I looked into it, the more I realized Mars doesn't look like the high Sierra — *Green* Mars looks like the high Sierra. Only with terraforming would I get that spectacularly beautiful and Alpine landscape that I had come to love.

So I thought, well, we have to terraform Mars, obviously, and there's no higher point to human existence. And on I went from there. At that point I had been writing novels for a number of years, I had had training in utopian thinking from Fredric Jameson, had learned from Gary Snyder about poetry and Buddhist practice, really I could not have been luckier in my teachers. And my own personal experience had had me out in a wilderness area that looked like the Mars in my mind, and could be made to look even more like it.

Then also I moved to a communally owned neighborhood while in the middle of writing the Mars books, which taught me a lot. Things kept happening, until it was like the moment Galileo called the *mirabili coniunctio*, the miraculous conjunction, everything coming together. There had never been the possibility of writing *the* terraforming Mars novel until then, and somebody who loved wilderness and who was thinking about and even living in alternative societies was, it seemed to me, the right person for the job.

It helped too that in the 1960s and 1970s history had seemed labile, almost explosive with possibility. History could go anywhere whatsoever; and yet suddenly in the 1980s and 1990s some kind of horrible lockdown occurred. Suddenly it seemed like history itself was massively entrenched and would never change again, that a giant capitalist world order had won and there was nothing any of us could do about it. And so for college students in particular, this became a very nice time to open up a novel that said there might be a chance for a completely clean slate, where we could start over again without all this baggage and without all these rules, and make a new world. Part of the miraculous conjunction was the *desire* for a terraformed Mars, the *desire* for a new start.

So I was lucky. And I wasn't the only one who was lucky. There is a whole group and generation of terraformed Mars novels from this period; but I deliberately did not read them. I didn't want to know, I didn't want to have any impositions on what I myself did. So I'm ignorant of the rest of the terraformed Mars novels, except for the quite beautiful epic poem called "Genesis" by the poet Frederick Turner, which I dared to read once I was into *Blue Mars*.

After that, and all too suddenly, there were scientists at NASA Johnson, in Houston, saying a Martian meteorite they had found appeared to have fossil bacteria in it. For me this felt like a disaster. If there is life on Mars now, whether it's bacteria or viral or whatever, we can't just go there and terraform it. It would be a scientific or moral monstrosity to discover other life in the universe — which would be probably the greatest scientific discovery of all time — and then go mess it up, or wreck it — destroy the lab, in effect. So life on Mars would complicate the terraforming project to the point where it might never happen, which would in effect change my novel from science fiction to fantasy. And I don't like fantasy. It takes a dead Mars to make my novel possible, so I'm rooting for that finding.

Fortunately for me, the nanobacteria that they thought they found are now considered to be too small to be alive in the sense we usually think of the word. There may be something very, very small that acts strange, but it's not alive, it's not a bacteria, as bacteria need to be bigger than those little fossils in the Mars meteorites.

I'm also much comforted by the words of Chris McKay, scientist at NASA Ames and my go-to guy when it comes to planetary science questions. He said that life probably, among other things, needs tectonic plate action; that if you had life bounce onto Mars inside a meteorite from Earth, or anywhere else, it had landed on a planet that wasn't being roiled at the tectonic level, so evolution wouldn't occur because things weren't being moved around. And since the average lifetime of a species is about ten million years and it has been about two *billion* years since Mars made a tectonic move — then, if there had ever been life on Mars at all, I could rest easy, because it has probably gone extinct at some point along the way, and the planet is now as dead as a doornail and ready for gardeners to move in. I like that idea.

This leads us finally, I suppose, past our moment, to the question of what next? What will Martian fiction writers write now? I imagine if I had to do it at all, which happily I don't, I'd begin thinking about robots, because really that's what we're sending to Mars. These rovers, if you proposed for them an "artificial intelligence" — not with consciousness, just algorithms, some quick but limited machine brain — and you then confronted this limited intelligence with an invasion of its planet, after that intelligence had been caretaking it for any number of hundreds of years, how would it react?

But any scenario like this is assuming humanity will manage to solve the problem that puts Mars off the table for the next century or two, which is global warming and climate change. Until we deal successfully with that, Mars is merely a footnote. If it can teach us more about global planetary management, ecological management, that would be a good thing and its only current justification. And we can make that argument, because accomplishments like finding the hole in the ozone layer came partly from studying the atmosphere of Venus. Certainly studying Mars has value for us, but it is not primary. Mars won't serve as a bolt hole or an escape hatch. It's crazy and immoral to say it could, and we have to remind people of that every time we discuss Mars from now on.

But a moment might come when we solve the problems of our current climate crisis, get robust, and go off to Mars in person. Then there might be robots already there that consider us to be an invasive species, a problem for the planet. Perhaps we will once again find there is a moment of resistance to the arriving humans — and aren't we then back to *The Martian Chronicles*, and the story of the ghosts resisting the arrival of the humans?

CHRONICLING MARTIANS*

Sha LaBare

First of all, let me admit that my title, "Chronicling Martians," is intentionally misleading. Traditionally, the word "Martian" is used in an anthropocentric sense: In common parlance, "Martian" evokes critters analogous to humans,[1] beings like ourselves who have language, technology, minds, civilization or whatever else we humans use to distinguish ourselves from the world-around ecology of other fauna and flora on Earth. In other words, the "Martian" reinforces rather than undermines anthropocentrism or human exceptionalism.[2] Indeed, the sleight-of-hand that allows "Martian" to stand for anthropomorphs[3] from Mars is itself analogous to the common usage of "Terran" or "Earthling" to include humans and exclude the equally Terran or Earthling other animals, plants, bacteria and critters on what we commonly refer to as "our" planet.

That "Martian," in popular usage, is a metonym for "alien" opens up questions about all depictions — and searches for — alien life. On the traditional view I'm briefly sketching here, both "Martian" and "alien" only ever stand for anthropomorphs. Indeed, just as "Martian" is a metonym for "alien," the process that makes "Martian" or "Terran" stand for an entire set of planetary ecological relations is metonymic as well. Associated with the Earth, humans earn the name "Terran." Alternatively, this is a synecdotic process, by which one part of Earth's ecology — humans — stands for the whole, as in the expression "all hands on deck." In either case, the common use of "Martian" and by extension "alien" is anthropocentric and anthropomorphic, excluding other Martians and other aliens from our awareness, just as our fellow critters on Earth are excluded from the prevailing anthropocentric view of life on this planet. Critiquing such traditional anthropomorphic and anthropocentric interpretations of the word "Martian" is my primary aim in this essay; I hope, in so doing, to open up this word — and the related word "alien" — to an ecological understanding.[4]

Many thanks to the organizers of the 2008 Eaton Science Fiction Conference, "Chronicling Mars," and to the many presenters whose talks heavily informed "Chronicling Martians" both as a presentation and in the current version. Specifically, I would like to acknowledge George Slusser, Terry Harpold, Howard Hendrix, David Hartwell, and Kim Stanley Robinson for their presentations, cited in this text.

The history of SF's romance with other planets is deeply marked by what Fredric Jameson calls "world reduction." In Jameson's view, world reduction serves, as in Ursula K. Le Guin's *The Left Hand of Darkness* (1969) and *The Dispossessed* (1974), to make humans and anthropomorphs appear alone in their worlds, placing in the foreground their behavior and intrigue while reducing the nonhuman world to a backdrop. Of course, nearly all mainstream[5] fiction employs world reduction in this sense, with stories unfolding as if humans were the only critters on Earth, and certainly the only critters that matter. If—as was frequently suggested at the "Chronicling Mars" conference— Mars and other planets are mirrors of our own, then it should hardly surprise us that world reduction holds as true on these planets as it does on ours; indeed, the ongoing extinction event on this planet can be seen as an exercise in applied world reduction, with the incredible diversity of life being "reduced" to make way for freeways, parking lots, and housing developments.

Of course, world reduction as a technique in SF worlding[6] dates back at least to Golden Age depictions of Venus and Mars as jungle and desert planets. Replete with such reduced worlds—with jungle, desert, and ice planets—the SF genre has generally ignored the facts of planetary ecology,[7] foregrounding anthropomorph behavior and, as often as not, metaphorizing human affairs through recourse to a planetwide pathetic fallacy.[8] It is as if the Occam's Razor of the imagination had been at work, making way for the simplest worlds possible; indeed, the supposed foregrounding of "worlds" in SF is arguably what makes Jameson's diagnosis of world reduction in Le Guin so telling. As Brian Stableford notes, the general lack of ecological awareness in SF shows up most clearly in the preponderance of dangerous carnivores in ecologies missing enough plants and herbivores to support them: Edgar Rice Burroughs's Barsoom, of course, immediately comes to mind (1993: 365).

But while world reduction may be a generalized problem in SF's relationship with ecology and other planets, at the same time it serves another function in SF, one which actually foregrounds worlding rather than backgrounding it. If, as Samuel R. Delany suggests, in SF the landscape[9] or world is itself the primary protagonist, while the episteme or worldview is the secondary one (1976: 333), then the very same process that turns nonhuman worlds into backdrops for human action also serves to foreground worlds as characters in their own right, agents who, by their very separation from human action can be seen to act on their own. Recalling Delany's insistence that SF is an object-oriented paraliterature rather than a subject-oriented literature (1994: 32), in SF we find that, in Donna Haraway's words, "the world [is] a witty agent and actor":

> Here is where science, science fantasy, and science fiction converge.... Perhaps our hopes for accountability, for politics, for ecofeminism, turn on revisioning the world as a coding trickster with whom we must learn to converse [1991: 201].

In the last forty years or so, ecological thinking has started to take hold both in the SF genre and in the wider mode of awareness I call, following Istvan Csicsery-Ronay, Jr., the SF mode.[10] Indeed, in light of SF's focus on worlds as agents in their own right, I argue that not only is ecology vital to the SF mode but also that the SF mode is vital to ecological thinking. Well-versed in worlding, thinking at global scales, imagining long time frames, and telling stories of first contact, SF is an important asset to ecology. Literally "household knowledge," the word "ecology" was first coined by Ernst Haeckel to draw attention to the relationship between organisms and their environments. My own understanding of ecology, however, eschews any easy dichotomy between critters and their surroundings. As in Donna Haraway's *When Species Meet* (2008), in my view ecology always implies that

> those who are to be in the world are constituted in intra- and interaction. The partners do not precede the meeting; species of all kinds, living and not, are consequent on the subject- and object-shaping dance of encounters [2008: 4].[11]

In other words, rather than focusing on the relationship between organism and environment, ecology in the sense I'm using here implies instead that we critters, as John Brunner might put it, actually *are* each other's environment.[12]

With this in mind, a brief discussion of ecology and humanity here on Earth is in order, as this planet provides the prototype for imagining a Mars, a Venus, or any other planet capable of supporting life. In fact, it was in the course of his work on planetary life detection — specifically, looking for atmosphere on Mars — that James Lovelock first developed Gaia theory. Until Gaia theory, the standard scientific — and SF — view was that planets simply had or did not have atmospheres permitting life; whether the entire planet was a desert, a ball of ice, or even (especially in *Star Wars*) a completely barren asteroid or moon made no difference. This tradition recalls, from Delany's *Dhalgren* (1975), Tak Loufer's "third convention" of the SF genre, that "the Universe is an essentially hospitable place, full of Earth-type planets where you can crash-land your spaceship and survive long enough to have an adventure" (415). Gaia theory, on the other hand, maintains that breathable atmospheres are a function of life and not vice versa — or rather, *and* vice versa. The biosphere, in Lovelock's view, might best be imagined as an enormous organism which assures, through world-around ecological homeostasis, the right environment for life, that is, for itself.

Among other visions, Gaia theory marks an early moment in a growing global ecological consciousness, one that in the latter half of the 20th and the first years of the 21st century is a vital, if not *the* vital, issue of our times. The threat of nuclear apocalypse has in recent years given way to a host of ecological catastrophes: rising oceans, megastorms, global plagues, overpopulation, and so on.[13] As global warming and mass species extinction may well attest, thinking ecologically — that is, thinking in harmony with this household called Earth — is essential for fruitful worlding, here on this planet and on any other.

A brief look at H.G. Wells's foundational Martian novel, *War of the Worlds* (1898), might clarify the intimate connection between SF and ecology. Wells's Martians invade Earth both to escape a dying Mars and to find, in humans, a new food source. While the ostensible focus of this war between worlds is a conflict between humans and Martians, the real war is an ecological one. Indeed, one implication of Wells's novel is that *any* war between worlds is an ecological war.[14] The Martian vines that signal the imminent areoforming of Earth are arguably a clue to this fact: Wells's Martians illustrate the perils of insufficient ecological thinking, allowing their alien kudzu to overwhelm the Earth while their invasive ecosystem is in turn consumed by Earthling bacteria.[15]

In depicting a Mars without bacteria, Wells is joined by Stanley G. Weinbaum, whose "A Martian Odyssey" (1934) and "Valley of Dreams" (1934) both take place on Mars.[16] This complete absence of bacteria on the Mars of Wells and Weinbaum raises the question of whether this is the mark of a truly alien world or simply an indication of world reduction. Having no bacteria at all, these Martian worlds are radically reduced compared to Earth, where life itself depends on these microscopic life-forms, a point argued at length by Lynn Margulis, the microbiologist who coauthored Gaia theory with James Lovelock. As Donna Haraway is fond of pointing out, the DNA in the space I call "my" body is only 10 percent human (2008: 3); the rest belongs to microbial symbionts who make human life — for example, human digestion — possible.[17] A Mars without bacteria — without microbes at all — is most certainly one in which ecological relations are radically simplified — if, indeed, life could exist at all in those conditions.

As Howard V. Hendrix suggests in his presentation on Mars and extremophilia, in the years since *War of the Worlds* bacteria have shifted from being signs of Martian death to being signs of Martian life. Olaf Stapledon's Martians, from *Last and First Men* (1930), might be seen as foreshadowing this shift, and, indeed, as critiquing Wells. Having their own areocentric view, these Martian microbial swarms consider us mindless inferiors and treat us the way we treat other terrestrial life-forms — that is, mostly ignoring us and getting on with their real business, which is waging war on radio towers and moving diamonds to the highest mountaintops.[18] Of course, the early SF tradition that depicted a Mars without bacteria has it backwards, much like Stapledon's Martians: We now know that if Martians exist at all, they *are* microbes, probably similar to Terran *archaea*, many of which are "extremophiles" preferring environments inimical to most other life on Earth. From an anthropocentric view, all microbes represent a kind of scalar extremophilia, recalling J.B.S. Haldane's suggestion that we humans are somehow "the right size," perfect intermediates between the largest and smallest scales in the universe. Of course, the very idea of extremophiles is anthropocentric, as if life itself where most comfortable at what humans call "room temperature." Mars, on this view, is an extreme planet, fit only for extremophiles.[19]

The notion that there are extreme "Martian" environments right here on Earth and that Mars is an extremophilic planet — judged as a deviation from the norm set by most if not all Terran life — illustrates, again, a not-so-subtle anthropocentrism. I mean this not just in the sense that Earth is a privileged planet (incidentally, the title of an intelligent-design propaganda film) but also that human worlds and worldviews are somehow privileged over other Terran venues for life. Indeed, if traditional anthropomorph Martians want to invade Earth so badly, this makes them less extremophiles than ... mediocrophiles? Right smack dab in the middle of a mass extinction event — only the sixth of its kind in the known history of our world — and this one caused in all likelihood by human civilization, in thinking Mars we should beware of how the notion of extremophilia might reinforce a lingering evolutionary fatalism, a now-classic SF obsession with survival of the fittest.[20] With life dying out all around us, will we humans be lucky enough, *fit* enough, to survive?

The meme of Mars as a dying planet and Martians as a dying race is of course directly derived from the work of Percival Lowell.[21] This tradition of the ancientness of both Mars and Martian civilization is accompanied by, as I suggested above, a planetary version of the pathetic fallacy, one which inevitably informs current understandings of Mars as a dying or lifeless planet.[22] Indeed, Lowell's vision of Mars arguably contributed to an evolutionist model of the inner planets in SF: Venus became a young, savage planet, a jungle planet swathed in swamps and forever a riot of rain and rot, while Mars became an old planet, a desert planet bereft of water and doddering into a decrepit, degenerate, and decadent death. In this evolutionist model, Earth, of course, is the privileged planet, containing aspects of the others in its jungles and deserts but otherwise providing the ideal environment for the pinnacle of life: modern humanity.

But not for long. Coupling the exhaustion of anthropomorph civilization with the exhaustion of the planetary ecosystem is, of course, hardly a farfetched notion. As a mirror for Earth, this traditional dying Mars offers an "if this goes on..." cautionary tale for humans on Earth, a planet which itself may, in an ironic twist, soon be in need of terraforming. Indeed, as Kim Stanley Robinson points out in his "Martian Musings," before we can even consider escaping to other planets, we must first prove ourselves by successfully dealing with the ecology of our own planet. Thinking ourselves and the other critters — all the other critters, extremophiles and mediocrophiles alike — as one single ecology is, I argue, vital to SF mode speculation, both here on Earth and on any other planet.

The critique of anthropocentrism and the promotion of ecological models of thinking are, I argue, key elements of the SF mode. While SF history indicates that much SF has been engaged in reinforcing human exceptionalism — that is, in policing the boundaries of the human — at the same time the mere evocation of other planets, other forms of life, and other ways of thinking undermines anthropocentric worldviews. Indeed, and thankfully, it is not in the nature of SF nor in the nature of SF criticism to offer final solutions, but instead to offer extended questionings, as-ifs, and thought-experiments. Evoking Teilhard de Chardin's distinction between the biosphere and the noosphere — the realm of biology and the realm of ideas or memes — Kim Stanley Robinson points out that on Mars, as on any planet considered a candidate for terraforming, the noosphere actually precedes the biosphere rather than vice versa.[23] Approaching varied and often contradictory visions of Mars and Martians as one ecology of memes, one noosphere, we can, I think, further affirm and encourage a tradition of ecological thinking in the SF mode.

In 1953, Michel Butor published a piece on SF and the "crisis of its growth," suggesting that SF writers collaborate on some single future world to keep their more farfetched imaginations in check. In some sense, the Martian noosphere we have discussed here at "Chronicling Mars" represents just such a single future world, and one which, like all worlds, is marked by contradiction, conflicting interests, and diversity.* Embracing this diversity — both in life-forms and memeforms — is, I think, the most ecological way to practice the SF mode. In conclusion, as Butor puts it, if instead of voyaging far out to the limits of space and time the SF writer had remained on Mars, he would have been obliged to invent something — something new, some new way of worlding to hold the many, the strange, and the alien together. The same, I hope, is true of Earth. If, as William Gibson once remarked, SF does not so much predict the future as colonize it, then understanding and critiquing the worldviews we use in our colonizing is a vital action in the now, one that opens up amazing possibilities — both here on Earth, and on Mars. Thinking Martian ecology means rethinking not only alien ecology in general, but also rethinking our own; perhaps some day soon we humans will make first contact with the bizarre other "kingdoms" of life here on Earth.

Notes

1. An important conceit of anthropocentric worlding is that the "human" remain a stable, even unmarked, category. Recent work in anti-humanist and posthumanist theory critiques this conception of the human, pointing to how the human is either defined through essential or primary qualities (e.g., the Christian soul, the Cartesian *cogito*, Hegelian reason) or — and perhaps more importantly — defined against its others, including not only other animals but also those humans who are considered somehow less than human: the slave, the woman, the person of color, the disabled, etc. Throughout my dissertation, *Farfetchings: On and in the SF Mode* (2010), I mark this critique typographically by replacing "human" with "?human," to recall that what and who gets to count as human remains always in question. Similarly, while the vague epithet "nonhuman" may be useful and even necessary for those of us concerned with anthropocentrism — e.g., SF studies, animal studies, postcolonial theory, ecofeminist ethics — it, too, must be marked: ?nonhuman. Given the scope of the current essay, I forgo such typographical markings here, although my word choices are always informed by this critique. The word "critter" provides an excellent case in point. This beautifully vague word sidesteps the dichotomy always already implied by the word "animal" (human/animal). Like the excellent word "widget," "critter" has the advantage of leaving very much open the kind of being we are dealing with.

2. For recent work on human exceptionalism, see especially Haraway 2008, Derrida 2008, Agamben 2004, and Wolfe 2003. My own thinking on this topic is less informed by the recent turn among continental philosophers than by work in ecofeminist philosophy and feminist science studies.

3. I am using "anthropomorph" here as a short form for "anthropomorphic aliens"; for further discussion see note 4.

4. This is not to say that anthropomorphism doesn't have its uses; indeed, anthropomorphism is arguably preferable to what Frans de Waal calls "anthropodenial" (1997), the anthropocentric tendency to deny commonality between humans and other animals. Anthropomorphs are primarily interesting in the way that they constitute research into the limits and boundaries of the humans — even, indeed, when they perform boundary work. In his excellent talk at "Chronicling Mars," Terry Harpold called attention to Jules Verne's critique of that strong form of

---

*Butor's piece was first published in English in 1967; I cite the Clareson because it is presumably the version my readers are most familiar with. In his "Remarks Occasioned by Dr. Plank's Essay 'Quixote's Mills'" (1973), Stanislaw Lem comments that Butor's dream has largely been accomplished, given the ways that SF functions inside of what Damien Broderick calls a "mega-text" (1995: 57–63).

anthropomorphism so familiar from contemporary SF television and film: "It does not do to dress up human beings in carnival attire, and call them Martians, or Moon Men, and it is this mistake which Mr. Wells so wonderfully and so successfully avoids. He invents his Moon Men and his Martians, and he gives them attributes which actual science really may permit them" (1928). Whether Wells actually avoids this error is open to question; in any case, at least, his aliens exploit a less obvious form of anthropomorphism than does *Star Trek* or *Star Wars*.

5. A problematic term, but I will leave it so. For insightful critique of the concept of the mainstream in SF criticism, see Luckhurst 1991 and 1994.

6. "Worlding" as a process is a concept I inherit from Donna Haraway. My own use of the term owes more to semiotician Thomas A. Sebeok and biologist Jakob von Uexküll than it does to Martin Heidegger. In 1963, Sebeok coined the term "zoosemiotics," borrowing and elaborating on Uexküll's "*Umwelt*," a term used to characterize the worldviews of different critters depending on their various sensory apparati. As ever, with the word "worlding" I find myself in the grips of what Le Guin calls "talking backwards" (Le Guin 1985), a subject I treat at length in my dissertation chapter of the same name. As with "human" and "nonhuman," discussed in my first note above, the words "world" and "worlding" simultaneously advance my argument while etymologically rooting me in a contradictory, even obsolete, worldview: Worlds — from the Germanic *wer-ald*, meaning "age of man"— are always already inhabited by something like man. Suffice it to say that "worlding" in the sense I'm giving it here need not be done by humans.

7. Frank Herbert's *Dune* (1965) is the most readily available counterexample; while the novel takes place on the eponymous desert planet, Herbert pays some attention to the — albeit highly reduced — ecology of this world, arguably itself an alternate Mars. Incidentally, in the subsequent series the planet Dune is terraformed, a process that kills the hydrophobic sand worms and creates an imperial monopoly on the psychedelic mélange, itself instrumental to faster-than-light travel. The parallels with our own most likely fatal addiction to oil are, of course, inescapable.

8. Also known as the anthropomorphic fallacy, the pathetic fallacy ascribes human motivations, feelings, etc. to nonhuman and usually inanimate objects. See note 4, above, for further discussion of anthropomorphism.

9. In an early draft of this essay, I used Delany's word "landscape" more extensively, but have opted for "world" in this version. On reading that draft, Sherryl Vint suggested that I avoid the word "landscape" because it connotes a passive backdrop, a sentiment echoed by Gwyneth Jones in her review of Le Guin's *Always Coming Home* (1985): "The feminised landscape of Utopia ... is a dangerous construction. In accepting this model feminist writers embrace an age-old tradition, in which 'woman' and landscape are one.... To put it succinctly: gardens do not write books. Landscapes, however beautiful, do not get up and read our critical papers at conferences. They may *act upon us* passively and gently, if we are receptive. They do not act" (203). As a third-wave critique of second-wave feminism, the point is taken; and yet, doesn't this critique further reinforce the sexist worldview it claims to undermine? The active/passive binary Jones invokes is far too complicit in the male/female binary for my tastes. I'm sure both Jones and Vint would agree that undermining this distinction might be a more interesting tack. It occurs to me therefore that I might have used "landscape" more extensively, in part to give agency to a concept often considered passive. We do not, *pace* Jones, need to be "receptive" for landscapes to act upon us; indeed, our current and coming ecological crisis should teach us that much.

10. Although Csicsery-Ronay, Jr.'s more recent work uses instead the term "science fictionality" (2008), I prefer the term "SF mode." I elaborate at length on SF as a mode of awareness and production in my dissertation, *Farfetchings: On and in the SF Mode* (2010). In that text, I argue that SF as a genre has structured our contemporary world-around ensemble, and in alignment with many of my colleagues in SF fandom and SF scholarship, I suggest that SF is best understood not as a genre — with its attendant border policing motivated by anxieties of legit-

imation — but instead as a mode available to SF agents across many disciplines, media, genres, and practices. The value of SF as a mode resides, not only in its ongoing critique of the world that is the case, but also in its contribution to creating what Sun Ra calls alterdestinies, that is, to conjuring forth alternate possibilities that call us into always already being differently in any given ecology, be it a classroom, a university, a state, a landscape, a biological or technological system, or any other context for human and nonhuman becoming.

11. Haraway here employs Karen Barad's concept of "intra-action": "The neologism 'intra-action' *signifies the mutual constitution of entangled agencies.* That is, in contrast to the usual 'interaction,' which assumes that there are separate individual agencies that precede their interaction, the notion of intra-action recognizes that distinct agencies do not precede, but rather emerge through, their intra-action. It is important to note that the 'distinct' agencies are only distinct in a relational, not an absolute, sense, that is, *agencies are only distinct in relation to their mutual entanglement; they don't exist as individual elements*" (2007: 33).

12. The reference is to John Brunner's *Stand on Zanzibar* (1968), in which the slogan "You are my environment and I am yours" (5) figures prominently; particularly influential people are said to be "environment-forming for all of us" (21).

13. For an exclusively filmic version of these familiar eco-catastrophes, consider in turn Emmerich's *The Day After Tomorrow* (2004), Spielberg's *Artificial Intelligence: A.I.* (2001), Boyle's *28 Days Later* (2002) and Fresnadillo's *28 Weeks Later* (2007), and Fleischer's *Soylent Green* (1973).

14. While this conclusion seems obvious, most readings of *War of the Worlds* focus instead on the issue of colonialism, which Wells himself explicitly foregrounds in the text.

15. The Martians — who survive by sucking the blood of others — are, as Howard Hendrix put it in his talk at "Chronicling Martians," literally "dead meat" on Earth; as such, they are consumed by Terran putrefactive bacteria. Mars, as Hendrix quips, is dead but doesn't know it yet.

16. Weinbaum is often — and rightly — praised for his "ecological" approach to Mars, but as one commentator at "Chronicling Martians" put it, his is more of a mish-mash than a fully realized ecology per se.

17. Consider, for example, this excellent passage from *When Species Meet*: "To be one is always to *become with* many. Some of these personal microscopic biota are dangerous to the me who is writing this sentence; they are held in check for now by the measures of the coordinated symphony of all the others, human cells and not, that make the conscious me possible. I love that when 'I' die, all these benign and dangerous symbionts will take over and use whatever is left of 'my' body, if only for a while, since 'we' are necessary to one another in real time" (Haraway 2008: 4).

18. That these Martians, after losing the war, still survive on Earth and enter into a symbiotic relationship with humans makes Stapledon seem even more prescient. For an extensive discussion of microbial symbiosis with humanity, see, for example, Lynn Margulis's *Symbiotic Planet* (1998).

19. In light of Gaia theory, a "lifeless" planet is necessarily one that is inimical to life, as in this passage from Greg Bear's *Moving Mars* (1993): "An unprotected human on the surface of Mars would very likely freeze within minutes, but first would die of exposure to the near-vacuum. If this unfortunate human survived freezing and low pressure, and found a supply of oxygen to breathe, she would still be endangered by high levels of radiation from the Sun and elsewhere.... After Earth, Mars is the most hospitable planet in the Solar System" (frontispiece, no page number). That Bear's novel takes place in the late 22nd century indicates the entrenched nature of anthropocentrism, indicating a preference for terraforming over the at-least-equally viable option of "pantropy," on which see especially Frederik Pohl's *Man Plus* (1976).

20. Wells and Stapledon aside, I'm thinking especially here of the novels of J.H. Rosny Aîné. In particular, George Slusser recalled to my attention Rosny's fascinating Mars novel, *Les Navigateurs de l'infini* (1925), in his talk on J.H. Rosny's Mars. In this novel, humans arrive on Mars

just in time to save the most anthropomorphic of the Martians from extinction, faced as they are with the emergence of two new — and more radically alien — kingdoms of life.

21. Percival Lowell, as Robinson argued in his "Martian Musings," should be seen as a bona fide SF writer — or, in my terms, "SF agent" — rather than as a hoaxer. As David Hartwell pointed out in his talk, Edgar Rice Burroughs borrowed his vision and maps of Mars from Lowell's first book, ignoring the later evolution of Lowell's vision; Hartwell also notes that Lowell's vision of Mars was debunked not long before the publication of his last book by Alfred Russel Wallace in his *Is Mars Habitable?* (1908). Hartwell does not, however, note that Wallace's critique was informed by his own anthropocentrism, which biased him to a teleological view of evolution and to the view that Earth was the only habitable planet in the universe (on which, see his *Man's Place in the Universe* [1902]). Both of these works are available at books.google.com.

22. In his "Martian Musings," Robinson suggests — perhaps a bit optimistically — that if life is discovered on Mars it will mark the planet as off-limits to terraforming. In his view, terraforming requires a truly "dead" planet; as one of the "Reds" from his Martian trilogy might comment, however, even the absence of life may not be sufficient justification for terraforming. Like David Brin — speaking with Greg Bear and Gregory Benford at "Chronicling Martians" — I'm fascinated with the human urge to breathe life into rock, to endow it with mind, but I hardly think that rocks, or rocky lifeless planets, are in need of "uplift." Indeed, in a very real sense the entire search for extraterrestrial life is always already anthropocentric: In most accounts, the tale of life, the tale of evolution itself is a teleological history leading inevitably to humanity. The search for life in the universe motivates quite a few interesting experiments and pushes the development of cool scientific techniques, but why are we so obsessed with it? What's so special about life that we would look for it everywhere? Why is life conceived as something that could "fill" an "empty" universe — especially given that we're quite busy wiping it out all around us right here on Earth? For an excellent piece on the more problematic aspects of the search for extraterrestrial life, see Stefan Helmreich's "The Signature of Life: Designing the Astrobiological Imagination" (2006).

23. The reference is to *Red Mars* (1993); similarly, in his "Martian Musings," Robinson elucidates on what he calls "Bradbury's Three Laws of Mars," the first of which is that the ghosts of our Martian stories will always be there.

Works Cited

Agamben, Giorgio. *The Open: Man and Animal.* Stanford, CA: Stanford University Press, 2004.
Barad, Karen. *Meeting the Universe Halfway: Quantum Physics and the Entanglement of Matter and Meaning.* Durham, NC: Duke University Press, 2007.
Bear, Greg. *Moving Mars.* New York: Tor, 1993.
Boyle, Danny, dir. *28 Days Later.* Fox Searchlight Pictures, 2002.
Broderick, Damien. *Reading by Starlight: Postmodern Science Fiction.* New York: Routledge, 1995.
Brunner, John. *Stand on Zanzibar.* New York: Ballantine Books, 1968.
Butor, Michel. *SF: The Other Side of Realism.* Ed. Thomas D. Clareson. Bowling Green, Ohio: Bowling Green University Popular Press, 1971.
Csicsery-Ronay, Istvan, Jr. *The Seven Beauties of Science Fiction.* Middletown, CT: Wesleyan University Press, 2008.
_____. "The SF of Theory: Baudrillard and Haraway." *Science Fiction Studies* 18: 3 (1991).
Delany, Samuel R. *Dhalgren.* New York: Bantam Books, 1975.
_____. "The Semiology of Silence: The *Science Fiction Studies* Interview." *Silent Interviews: On Language, Race, Sex, Science Fiction, and Some Comics.* Wesleyan University Press, 1994.
_____. *Triton.* New York: Bantam Books, 1976.

Derrida, Jacques. *The Animal That Therefore I Am.* Ed. Marie-Louise Mallet, Trans. David Wills. New York: Fordham University Press, 2008.
Emmerich, Roland, dir. *The Day After Tomorrow.* 20th Century–Fox, 2004.
Fleischer, Richard, dir. *Soylent Green.* Metro-Goldwyn-Mayer, 1973.
Fresnadillo, Juan Carlos, dir. *28 Weeks Later.* Fox Atomic, 2007.
Haldane, J.B.S. "On Being the Right Size" (1928). *On Being the Right Size and Other Essays.* Ed. John Maynard Smith. Oxford University Press, 1985.
Haraway, Donna. "Situated Knowledges: The Science Question in Feminism and the Privilege of Partial Perspective." *Simians, Cyborgs, and Women: The Reinvention of Nature.* New York: Routledge, 1991.
_____. *When Species Meet.* Minneapolis: University of Minnesota Press, 2008.
Helmreich, Stefan. "The Signature of Life: Designing the Astrobiological Imagination." *Grey Room* 23: 4 (2006).
Jameson, Fredric. "World Reduction in Le Guin: The Emergence of Utopian Narrative." *Science Fiction Studies* 2: 3 (1975).
Jones, Gwyneth. "No Man's Land: Feminised Landscapes in the Utopian Fiction of Ursula Le Guin." *Deconstructing the Starships: Science, Fiction and Reality.* Liverpool University Press, 1999.
Le Guin, Ursula K. *Always Coming Home.* 1985. Reprint, Berkeley: University of California Press, 2001.
_____. *The Dispossessed.* New York: Avon, 1974.
_____. *The Left Hand of Darkness.* 1969. Reprint, New York: Ace, 1987.
Lem, Stanislaw. "Remarks Occasioned by Dr. Plank's Essay 'Quixote's Mills.'" *Science Fiction Studies* 1: 2 (1973).
Luckhurst, Roger. "Border Policing: Science Fiction and Postmodernism." *Science Fiction Studies* 18: 3 (1991).
_____. "The Many Deaths of Science Fiction: A Polemic." *Science Fiction Studies* 21: 1 (1994).
Margulis, Lynn. *Symbiotic Planet: A New Look at Evolution.* New York: Basic Books, 1998.
Robinson, Kim Stanley. *Red Mars.* New York: Bantam Books, 1993.
Rosny Aîné, J.H. *Les Navigateurs de l'infini.* Paris: Présence du Futur no. 25, 1983.
Spielberg, Steven, dir. *AI: Artificial Intelligence.* Warner Bros. Pictures, 2001.
Stapledon, Olaf. *Last and First Men: A Story of the Near and Far Future.* London: Penguin Books, 1937.
Verne, Jules. "Solution of Mind Problems by the Imagination" (1928). http://www.julesverne.ca/vernebooks/jvmindproblems.html.
Waal, Frans de. "Are We in Anthropodenial?" *Discover* 18: 7 (1997).
Weinbaum, Stanley G. *The Best of Stanley G. Weinbaum.* New York: Ballantine, 1974.
Wells, H.G. *The War of the Worlds.* In *Three Novels of the Future.* Garden City, NY: Nelson Doubleday, 1979.
Wolfe, Cary. *Animal Rites: American Culture, the Discourse of Species, and Posthumanist Theory.* Chicago: University of Chicago Press, 2003.

Three

Science and Fictional Mars

MARS AS CULTURAL MIRROR
Martian Fictions in the Early Space Age
Robert Crossley

There have always been writers whose visions of Mars were oblivious to scientific advances in the study of the planet, if not resolutely counterfactual. But perhaps the most uneasy chapter in the cultural history of Mars occupies the years between the early space launches of the 1950s and the processing of *Mariner 9*'s photographs in 1972. In 1950 Ray Bradbury, in the last flowering of Romantic Mars, could still write sumptuously about an impossible Mars, but in the years that followed the publication of *The Martian Chronicles* there was no new, authentically science-fictional model to supplant it. Throughout the 1950s discoveries about Martian atmosphere, temperature, and topography accumulated, and at the time of the 1956 Opposition, a year before the launch of *Sputnik*, astronomers were focusing on the still unsettled question of whether simple lichens and mosses survived on the surface on Mars. Larger life-forms, including animals, had already been ruled out by most reputable astronomers. An editorial in a Boston newspaper regretfully advised readers not to get caught up in the scientists' excitement over the Opposition: "Put your telescope on Mars and you'll ruin a romantic dream. Watch television these nights instead of the southeast sky and you can continue to think in terms of Buck Rogers, flying saucers and beautiful Martian girls" ("Mars on Trial," 20).

The director of Philadelphia's Fels Planetarium noted that 1956 was also the fiftieth anniversary of the publication of Percival Lowell's *Mars and Its Canals*. An era was ending, he noted. The "simple, unadorned telescope" on which Schiaparelli and Lowell had depended was now supplemented by a new arsenal of tools: spectroscopy, radiometry, photoelectric photometry. And, he added, the world was on the verge of the physical exploration of space which would soon eclipse the technology of the terrestrial telescope (Levitt, vii–viii). As the infant space programs in both the Soviet Union and the United States worked toward the first probes of the Red Planet in the 1960s, the gulf that had existed between the scientific and the literary imaginations ever since the Lowell controversy grew even wider than before. Mars became the holy grail for planetary astronomers once the possibility of sending cameras into space was achieved, but it became an onus for the imaginative writer. One significant difference from earlier periods is that by the close of the 1950s the general public had been weaned away from Lowellian fantasies. People no longer formed their impressions of Mars from Sunday supplement articles but from the new popular medium of television which, despite the Boston editorialist's pronouncement, would contribute to the debunking of the Mars of legend. In 1954 ABC began broadcasting *Disneyland*, a television anthology divided into the four segments of Adventureland, Frontierland, Fantasyland, and Tomorrowland. One of the highlights of the 1957 season was Disney's "Mars and Beyond," a "science-factual" documentary to which Wernher von Braun had been a consultant. Using animation and simulation techniques, along with talking-heads commentaries, "Mars and Beyond" offered, as one journal of astronautics phrased it, "a sober view of contemporary scientific hypothesis and conjecture" ("Space Preview," 29). *Time* magazine commended the Disney writers: "They did not confuse the popular with the vulgar, avoided the error of talking down to the viewer" ("Television: Review"). Similarly, a new magazine, *Space Age*, reported early in 1959 that the Air Force (NASA, the National Aeronautics and Space Administration, had only just been authorized by Congress) was developing a program for sending scientific probes to Mars. But the writer cautioned against unfounded expectations of sensational revelations. His article opens with the astringent assertion, "Mars is a graveyard" (Caidin, 9).

In the midst of this new sobriety, fiction about Mars didn't quite die out, but it often took odd forms. Frank Herbert gave readers a Mars-like planet in *Dune* with a parched, sandy desert, breathable atmosphere, and exotic fauna out of the Burroughs menagerie. As Oliver Morton has observed, when Baron Harkonnen points to a globe of the planet Arrakis with its small polar caps, caramel-colored terrain, and absence of oceans or rivers there is no doubt what planet is being referenced. (Herbert, 20–21, 523; Morton, 178). To get away with creating a literary Mars in the Lowell-Burroughs mold, Herbert had to move it out of our solar system. Other writers had their own strategies for smuggling the old Mars into new dustjackets, but many Martian narratives of the 1960s became more mythic and reflexive in method, declaring themselves artifices, playfully quoting not only from the literature of Mars but from the broader literary tradition, and often veering toward either parable or parody. About the Mars being revealed by science they were frequently evasive or silent, and sometimes perversely retrograde.

Back in 1941 the *New York Times* had written the obituary for one of the commonest of post–Schiaparelli Martian fantasies. "He was a useful vehicle, the Man from Mars, and now he is no more" ("Topics," 20). But if no one could quite believe in Martians any more, two influential works of fiction of the early 1960s dusted off the retrograde figure of the Man from Mars as "a useful vehicle" for ironic mythmaking. Valentine Michael Smith, the titular *Stranger in a Strange Land* by Robert Heinlein, and T.J. Newton in Walter Tevis' *The Man Who Fell to Earth* are Martian exiles on Earth, the former a Prometheus figure and the latter, as the novel's title advertises, a contemporary

Icarus. In the tradition of the political and utopian allegorists of the 1890s, Heinlein and Tevis use men from Mars as devices for achieving a subversive vision of American society — a send-up of American sexual puritanism and fundamentalism in *Stranger in a Strange Land*, and in *The Man Who Fell to Earth* a rather darker and more probing examination of political chicanery and cultural spoilage. Heinlein's Mars, glimpsed only briefly at the beginning of the novel, has the old apparatus of canals, ruins, and an ancient race of savants. Smith, born to a couple on the first terrestrial expedition to Mars but raised by indigenous Martians, travels back to his native world where he is both an alien and a messianic presence. His status as prophet without honor, man without a planet, and sexual hedonist without inhibition had much to do with the novel's adoption as a countercultural talisman in the later 1960s, but his identity as Martian is an antiscientific gesture to literary tradition.

Tevis' *The Man Who Fell to Earth* is better fiction, elegiac and moving. The utopian Martian visitor who had become a cliché in fiction and drama is given a valedictory and distinctly postmodern treatment. Mars is never specified as the home planet of the alien who disguises himself as a Kentuckian and calls himself T.J. Newton, but there are persistent suggestions that the planet Anthea — located in our solar system, dying of thirst, its minerals and fuels nearly exhausted, its population reduced to a fading remnant of a vibrant and advanced civilization — is the Mars of popular lore. A curious Midwestern scientist spends much of the novel trying to decide whether Newton's strangeness means that he's from Mars or Massachusetts. When he finally has physiological evidence that Newton cannot be *Homo sapiens*, he asks him, "Is it Mars?" (Tevis, 129) But Newton doesn't respond — as neither he nor Tevis ever does to explicit questions about the identity of Anthea.[1]

The degeneration of Newton into a lethargic, genteelly drunken, and, finally, blinded imitation of a man points up Tevis' concern to use his Martian as both an observer of and a participant in the culture of a degenerate America. The Martian visitor experiences and exposes corporate greed, academic pretensions, governmental corruption, the lazy-mindedness of American consumers, and the sheer witlessness and incompetence of the large bureaucracies that shape people's lives. Framing his narrative with the ancient technological myth of Daedalus and Icarus, Tevis makes his tale of an illegal alien from Mars a parable about the potential for human liberation or disaster in the inventions of applied science. *The Man Who Fell to Earth* was first published in the wake of a flood of nuclear nightmare fictions that included Nevil Shute's *On the Beach*, Walter Miller's *A Canticle for Leibowitz*, and Mordecai Roshwald's *Level 7*, and in certain respects, Tevis' novel is even more despairing than those grim depictions of extinction. T.J. Newton, weeping and drunk on gin and bitters at the novel's end, has decided that the Earth is so corrupt, its civilization so subhuman that it is not worth saving. Tevis reprises the misanthropic mood of H.G. Wells's final (1937) novel about Mars, *Star Begotten*, in which "the superman makes an aeroplane, and the ape gets hold of it" (Wells, *Star Begotten*, 110). In fact, Tevis rediscovered precisely what Wells had figured out sixty-five years earlier in *The War of the Worlds*: that a central value of visitors from Mars as fictional devices is to hold the mirror up to human nature. "What are these Martians?" Wells's hysterical curate asks during the invasion. "What are we?" the narrator replies, reminding his late Victorian readers of the ruthlessness of Britain as an occupying global power (Wells, *War of the Worlds*, 103).

Philip K. Dick produced two novels with Martian locales in 1964, *The Three Stigmata of Palmer Eldritch* and *Martian Time-Slip*. Both depict a near-future Mars, colonized by terrestrials and with the features of a frontier society: "a new place" where people could start "a new life (Dick, *Martian Time-Slip*, 24), but also a dreary environment full of "half-abandoned gardens and fully abandoned equipment" and "great heaps of rotting supplies" (Dick, *Three Stigmata*, 121). Dick draws on standard prespace age iconography, including canals, a breathable atmosphere, and an ancient civilization in decline. The few surviving indigenous, dark-skinned Martians are nomadic and impoverished, as marginalized and exploited by the terrestrial colonists as they might have been by the scruffiest of Bradbury's American settlers on Mars.

If Dick's images of Mars are retrograde, it is also true that scientific plausibility was no more a priority for him than it had been for Bradbury. Even Dick, however, could not ignore the unprecedented photographic images of Mars that began to enter the collective consciousness when *Mariner 4* with its camera flew by the planet in 1965 and sent back 22 grainy pictures of a heavily cratered and apparently lifeless world. These first close-up views of Martian topography were profoundly shocking. Scientists wanted to put the canal question to rest but they still had been hoping for colorful signs of vegetation and other Earth-like features; instead, the Mars they saw was a blitzed and pulverized wasteland, inhospitable to both biological life and the life of the imagination. The lead engineer assigned to transform the graphic data sent back by *Mariner 4* into functional pictures told his assistant, "You've just seen the death of the planetary program" (quoted in Hanlon, 110). The images of a dead Mars were just as threatening to literary production. A year after *Mariner 4*, in "We Can Remember It for You Wholesale," Dick gave up the fantasy of hominid Martians and a breathable atmosphere, and settled for the still-conceivable possibility of worms and cactus-like vegetation. Mars, with its landscape of "profound gaping craters," is a "world of dust" where a space traveler from Earth is constantly "checking and rechecking [his] portable oxygen source" (Dick, "We Can Remember," 110–111). The uneager transition to a new literary Mars was beginning.

Novelists less famous than Heinlein, Tevis, or Dick can illustrate something of the dilemma of Martian fiction at the historical moment when old mythologies had not yet been fully displaced by the new views of Mars from the first space probes of the mid–1960s. Let me mention just one. D.G. Compton in *Farewell, Earth's Bliss*, first published in 1966, imagines Mars as a dumping ground for Earth's political, sexual, and criminal undesirables who are given a one-way ticket to the planet. It is an ugly world in every respect, as each new annual shipload of convicts arrives and surveys the empty and arid scene before them, although the newly arrived are advised sardonically by old-timers to think of Mars as a delightfully sunny environment with sweeping views and a thrilling sense of distance. "We believe in the beauties of our landscape," one of them warns some newcomers. "You're advised to think twice before making derogatory remarks about it" (Compton, 60). There in a nutshell is the artist's dilemma about Mars in the 1960s, torn between the desire to believe in the beauty of the planet and the unwelcome new evidence of the camera's eye. Compton's Mars is a cul-de-sac, and the colonists develop their own crude, reactionary, and malignant system of governance. Like Wells, Compton suggests a parody of British colonialism in the self-important speech of the settlement's "Governor," who insists he is "not another of those pathetic Englishmen changing for dinner and drinking cold tea out of sherry glasses on some rotting verandah in the Burmese jungle" (133). But the strength of *Farewell, Earth's Bliss* is not so much in its political parable as in its evocation of an existential despair and brutishness that this inhospitable planet calls forth from its human occupants.

Other novelists of the 1960s were more reluctant to give up the trailing clouds of glory of Lowellian and Bradburyan Mars. In "Tomorrow's World," a preface to his 1962 novel *Marooned on Mars*, Lester del Rey lamented that there was a time when scientists were convinced of the presence of intelligence on Mars. Now, faith in a geometrical system of canals has largely been lost and the likelihood of a Martian civilization past or present has dwindled, but "the markings on Mars are real" and the presence of vegetation plain to see through the telescope, del Rey claims. But he hedged his language with uncertainties: "Probably the canals are only some natural phenomena which have nothing to do with intelligent life" (del Rey, 6–7). Certainty would be achieved only when we travel to Mars to see the planet for ourselves, and the fiction that follows this preface imagines the first such voyage. "The technical details are generally accurate," the author announces confidently, "and nothing here is really fantastic" (7).

Like the Disney television show, *Marooned on Mars* claimed to be "science-factual." Once del Rey's narrative begins, however, the limitations of his realism become all too evident when the spaceship's pilot and passengers indulge in coffee and banana cream pie just a few minutes before they climb up a ladder and flop onto mattresses for the launch. *Marooned on Mars* is an ambivalent and not very successful effort to reconcile the pleasures of romance with the progress of planetary science. Chapter 12, titled "The Mysterious Canals," is particularly revealing of del Rey's hesitancy as a fiction writer to give up the old Mars. The crew had been charged specifically to solve the mystery of Lowell's canals, and what they discover are rows of pumpkin-like green vegetables linked by tubular filaments that convey from plant to plant the scarce water from melting polar caps. It remains unclear whether these "canals" are the product of agriculture or of vegetative adaptation to a desiccating environment, of manipulative intelligence or natural selection. The similes used to describe the appearance and function of the filaments — they are like pumps, like laundry lines, like telephone poles — imply intelligent design, although the only Martians the expedition discovers are three-foot tall bipeds who live underground as a refuge from the cold and who appear never to have reached the threshold of industrialization before their decline into savagery.

In the Early Space Age, fiction about Mars seemed to have nowhere to go. In 1971 the editors of a selection of a hundred years' worth of writing about Mars catalogued the visual images and the names of astronomers and fiction writers who had established the tradition of literary Mars. They gave the anthology the nostalgic title *Mars, We Love*

You. The editors knew the old imagery could no longer be replicated, but they insisted on the invulnerability of literary history to scientific advancement. Of Wells, Burroughs, Heinlein, Schiaparelli, and Lowell they wrote, "Taken as a whole, they have made of our nearest planetary neighbor a modern myth that will endure even when future interplanetary probes return pictures clearer than those of *Mariner 4*" (Hipolito and McNelly, ix). That anthology was barely into print when exactly those clearer photographs of Mars, snapped by *Mariner 9*, started being processed at the Jet Propulsion Laboratory in Pasadena. But before I turn to *Mariner 9* there is one last literary text I want to put before you.

Among the various resistant and retrogressive literary responses to the new Mars, one book responded in earnest to the *Mariner* era. Of all the fiction about Mars published between 1956 and 1976 I know of none that takes the current state of knowledge about the planet more seriously than *The Earth Is Near* by the Czech writer and painter Ludek Pesek. Published in German in 1970, it was translated into English in 1973. Powerfully influenced by the *Mariner 4, 6,* and *7* photographs, its view of Mars is probably grimmer than it would have been if Pesek had seen the more varied landscapes disclosed by *Mariner 9* in 1972. Nevertheless, his book rests on a deeper literary and scientific commitment to fiction-making about Mars than anything else in the period. It is so relentlessly anti-romantic in its approach that it makes a stark contrast with the regressive imagery, style, and concepts of many other writers.

The purpose of Pesek's imagined "Project Alpha" is to settle definitively the question of whether any life, including micro-organic life, exists on Mars. The biological experiments the novel recounts answer the question roundly in the negative. In the process the planet is revealed as a hostile, even nauseating environment for human beings. "God, I feel like death!" says the first crew member to stagger to a porthole and view the alien landscape (Pesek, 73). The "monotonous desert" and "shallow craters" (79), the "dead stone" (151), the landscape "reminiscent of the rock formations we knew from the Moon" (144)—all derived from the sterile images conveyed by *Mariner 4*—are matched psychologically by a sense of unrewarding tedium and futility in the crew's experience. The mechanical probes had prepared the crew for a planet that would not be homelike and for the unlikelihood of any complex life-forms. "Yet each of us," the narrator admits, "cherished some of that sweet, irrational romanticism that will overstep the bounds of probability" (122). Pesek gratifies none of those cherished desires. "An expedition to Mars is not a dream," the narrator later writes, "but a life and death struggle" (156). The ubiquitous dust causes far more problems than anticipated, fouling filters, blowing into pressure chambers, seeping into clothing, impairing scientific instruments, disabling tractor vehicles. Spending so much time in suits and helmets isolates crew members from one another and leaves them unable to see another's eyes and expressions, deadening the emotions and the sense of community. The nomenclature of the Schiaparelli maps, which for a century had turned Mars into a fantasyland, loses all its poetry. "Dusty desert stretched as far as the eye could see, a hollow inventory of names that had once sounded so romantic—Aeria and Arabia, and Moab and Edom on the other side. They dried up our throats, they crunched between our teeth like sand" (193). The form of the narrative complements the nondescript landscape and the loss of affect; it is a log of events, deliberately unadorned and undramatic, with repeated descriptions of failed equipment, failed experiments, abridged hopes, journeys that go nowhere.

The climactic episode in this mostly anticlimactic adventure on Mars centers on a biological survey team making a final 150-day "great march" in a last, and unsuccessful, effort to collect some evidence of life. But the real key to Pesek's purposes is in a lesser episode that interrupts the great march when a geological survey team reports back that it has located the ruins of a city in the desert. Weather conditions prevent them from getting close enough to collect a brick to bring back to the base for study, but "staggering pictures" are taken with a telephoto lens — photos that anger the head of the biological survey team for the sheer absurdity of the geologists' claim. The narrator writes on behalf of all the biologists on the expedition, "I don't think any of us were inclined to believe fairy tales of some lost Martian civilization" (157). When the site of the supposed city is finally checked out it proves to be nothing more than "weathered rocks buried in drifts of sand" (160). Comparison of new photos with the original telephoto lens pictures reveals illusions created by the refraction of light in the dust-laden atmosphere. Reminiscent of the optical illusion of Lowell's canals, the Martian city quickly fades into nonsense, and Pesek again refuses to take the romantic turn from which so many authors of Mars narratives had been unable to abstain.

The Earth Is Near ends in failure and death and unfulfilled obsessions, and Mars itself triumphs over the ambitions of the terrestrial expedition, blowing away the tracks of the explorers, remaining icy, hostile, silent, immutable in the face of the human incursion. The narrator reflects at the end that if space exploration is to have a future human beings will have to find a way to change themselves — their blood chemistry, their breathing, their body temperature, their mental habits — rather than haul their artifices and supplies and preconceptions with them. They must adapt to new worlds rather than expect those worlds to become tractable to human needs. Maybe Pesek would have altered some of the topographical details of his novel had he seen the *Mariner 9* images of Mars, but the visionary center of his novel — its acknowledgment of the otherness of Mars, its cautions about the seductiveness and perils of romantic dreaming, and its radical revision of the tradition of Martian adventure fiction — is powerful and original. As a portrait of Mars and an account of the nature of planetary exploration it is the strongest piece of fiction to appear between Bradbury's *Martian Chronicles* in 1950 and Kim Stanley Robinson's *Red Mars* in 1993.

The first space age revolution in images of Mars was the product of *Mariners 4* through *7*, whose photographs of a heavily cratered and crumpled marscape evoked a monotonous, barren, and nearly featureless world. But a second revolution occurred in the 1970s. *Mariners 8* and *9* were launched in 1971 and were fundamentally different in nature and mission from the earlier probes that had photographed localized portions of Mars as they briefly sped by the planet before passing deeper into the solar system. These two would go into orbit around Mars and were expected to provide months of photographs, collecting images of every region of the planet. *Mariner 8* failed at launch, but *Mariner 9* was reprogrammed to attempt to cover nearly all the experimental and photographic objectives of both missions. Late in 1971 *Mariner 9* rendezvoused with Mars. The anticipation that this encounter would result in the most extensive and rewarding haul of Martian data yet prompted a symposium at the California Institute of Technology (Caltech), home to the Jet Propulsion Laboratory where the data from *Mariner 9* would be received and processed. The symposium was a small gathering but a distinguished one: two scientists, two imaginative writers, and a science journalist. The host was Bruce Murray, astrogeologist on the Caltech faculty and the laboratory's director, and the panel he assembled included Cornell astronomer Carl Sagan, writers Ray Bradbury and Arthur C. Clarke, and *New York Times* science correspondent Walter Sullivan. They met on the evening before *Mariner 9* was due to go into orbit around Mars, and their discussion comprised equal parts of speculation about the new knowledge they would be gaining and homages to the old mythology of Mars. If one were looking for a distinct boundary line in the literary history of Mars between the old traditions and the new scientifically based fictions it could be drawn through the *Mars and the Mind of Man* symposium at the Jet Propulsion Laboratory on November 12, 1971.

Had Martian topography actually been visible on November 12, perhaps the symposium would have focused on the first new pictures from *Mariner 9*. But Mars was enveloped in a planet-wide dust storm that frustrated the impatient scientific team at Caltech. Instead, the panelists mixed pungent and sometimes comic reflections on the follies of past speculations about Mars with fond acknowledgments of the influence of even the most absurd and discredited theories on the participants' careers. Sagan rehearsed the obsessions of the 1890s, including Samuel Phelps Leland's prediction that the University of Chicago's newest telescope would soon disclose views of Martian cities, harbors, navies, and factory smoke (Bradbury et al., 14). Sullivan, the oldest member of the panel, recalled the 1924 Opposition — the closest of the twentieth century — when the director of the U.S. Army Signal Corps and the chief of Naval Operations ordered that silence be maintained at their radio stations in order better to track any possible radio signals being sent from Mars to Earth (6). Murray talked about the continuing susceptibility of the scientific mind, not just the popular imagination, to wishful thinking about Mars. The imaginative writings of Lowell, Burroughs, Bradbury, and Clarke, Murray suggested, instilled a desire for Mars in the general public that made them willing to pay for missions like *Mariner 9*, but desire itself thwarts the capacity to see "the real Mars." "We *want* Mars to be like Earth," he said in explaining why the propensity to construe data so as to support the hypothesis of Martian life had continued even as recently as the *Mariner 6* and *7* fly-bys in 1969:

> There was a misinterpretation of the spectral results from one of the instruments on board initially because, I feel, the person really wanted to believe that he discovered something that was a real clue to the existence of life on Mars. In fact, he had found something else that was extremely important, which indicated that parts of the Martian polar caps were not just CO_2 generally but absolutely pure dry CO_2 with no moisture at all upon them. He had made a very important discovery. But he had initially misread it, I feel, because of the expectation of seeing something else [22–23].

The November 12 symposium was preoccupied with the myth and the romance of Mars, and Bradbury was its chief spokesman. He argued that it was fundamental not only to human nature but to the process of scientific discovery "to start with romance and build to a reality" (35). Bradbury jokingly played to the audience with the fantasy that when the current dust storm passed and the cameras got a clear view of the planet they would see Martians carrying signs reading "BRADBURY WAS RIGHT!" (19). But even he had to admit that they were entering a new era and that his own Martian stories were now being taught in schools as a form of modern mythology. Once the Martian storm subsided and pictures began flowing back to Earth it would become apparent that while the old myths were indeed definitively inoperable, the new Mars was not at all what people had been led to suspect from the earlier *Mariner* probes.

Only the first third of *Mars and the Mind of Man* is taken up with the transcript of the November 12 symposium. The remainder of the book is composed of reflective essays by each of the participants more than a year later, after *Mariner 9*'s stunning 7,500 photos had been received and assimilated into a revolutionary revision of the scientific understanding of Mars. *Mariner 9* revealed the never-before-seen gigantic volcanoes of Mars — the most titanic mountains in our solar system — and smooth plains with sinuous channels that suggested the work of floodwaters in a much earlier wet period of the planet and, most dramatically, Mars's vast equatorial canyon system, 3,000 miles in length and four times the depth of the terrestrial Grand Canyon. Here was a landscape that could once again engage the romantic imagination, even if it lacked so much as a fungus or a lichen. Sagan, in his retrospective piece, emphasized that the vegetation theory that had so long been appealed to as an explanation for the seasonal darkenings of the surface observed through terrestrial telescopes was conclusively dead. With the new understanding of how gigantic storms rearrange the sandy fines on the planet and cause variations in reflected light on the surface, meteorology and geology began to seem more crucial than the life sciences for unlocking the secrets of Mars. When the *Viking* Landers reached Mars in 1976, Sagan predicted, organic chemistry experiments would not necessarily be more important than analyses of minerals, atmospheric gases, volcanic and seismic activity, and wind speeds and patterns. In his essay Clarke described the transcript of the 1971 panel discussion a year earlier as a relic of "the prehistory of Martian studies" (79). Envying the writers whose careers would be founded on the post–*Mariner 9* Mars, Clarke was confident that science would enhance rather than diminish the "magic" of Mars and provide for fiction "far more strange and wonderful than the wildest fantasy" (85). Sullivan echoed Clarke, pronouncing the obituary for the "dreamworld" of old Mars (117) but insisting that "no myth or legend could be as rich in beauty, wonder, and awe as the full reality of the universe that is our home" (127). The "real Mars" to which Murray and the Caltech scientists aspired turned out to be so geologically dramatic that it was capable of engendering a new romance with the planet. The stage was set for the passing of that retrograde phase of Martian fiction that had dominated writing for two decades and the launch of a new realism in the representation of Mars.

Note

1. Tevis, more interested in mythical resonances than astronomical science, has not troubled to provide reliable clues about the identity of Anthea. A reference to 35 Anthean years being the equivalent of 45 terrestrial years, for instance, leads the reader on a wild goose chase, since no planet in the solar system fits that equivalency.

Works Cited

Bradbury, Ray. *The Martian Chronicles*. New York: Doubleday, 1950.
Bradbury, Ray, Arthur C. Clarke, Bruce Murray, Carl Sagan, and Walter Sullivan. *Mars and the Mind of Man*. New York: Harper and Row, 1973.
Caidin, Martin. "The Mars Probe." *Space Age* 1 (February 1959).
Compton, D.G. *Farewell, Earth's Bliss*. 1966. Reprint, San Bernardino, CA: Borgo, 1979.
Del Rey, Lester. *Marooned on Mars*. 1962. Reprint, New York: Paperback Library, 1967.
Dick, Philip K. *Martian Time-Slip*. New York: Ballantine, 1964.
_____. *The Three Stigmata of Palmer Eldritch*. 1964. New York: DAW, 1983.
_____. "We Can Remember It for You Wholesale." 1966. Ed. Gordon van Gelder. *Fourth Planet from the Sun: Tales of Mars from The Magazine of Fantasy & Science Fiction*. New York: Thunder's Mouth, 2005.
Hanlon, Michael. *The Real Mars*. New York: Carroll and Graf, 2004.
Heinlein, Robert. *Stranger in a Strange Land*. New York: Putnam, 1961.
Herbert, Frank. *Dune*. New York: Ace, 1965.
Hipolito, Jane, and Willis McNelly, eds. *Mars, We Love You: Tales of Mars, Men and Martians*. New York: Pyramid, 1973.
Leland, Samuel Phelps. *World Making: A Scientific Explanation of the Birth, Growth and Death of Worlds*. Chicago: Woman's Temperance Publishing Association, 1896.
Levitt, I.M. *A Space Traveler's Guide to Mars*. New York: Henry Holt, 1956.
Lowell, Percival. *Mars and Its Canals*. New York: Macmillan, 1906.
"Mars on Trial." *Boston Herald*, September 7, 1956.
Miller, Walter. *A Canticle for Leibowitz*. Philadelphia: Lippincott, 1960.
Morton, Oliver. *Mapping Mars: Science, Imagination, and the Birth of a World*. New York: Picador, 2002.
Pesek, Ludek. *The Earth Is Near*. Trans. Anthea Bell. Scarsdale, NY: Bradbury Press, 1973.
Robinson, Kim Stanley. *Red Mars*. New York: HarperCollins, 1993.
Roshwald, Mordecai. *Level 7*. 1959. Ed. David Seed. Madison: University of Wisconsin Press, 2004.
Shute, Nevil. *On the Beach*. London: Heinemann, 1957.
"Space Preview: Mars and Beyond." *Space Journal* 1 (Spring 1958).
"Television: Review." Disneyland: "Mars and Beyond." *Time* 9 December 1957.
Tevis, Walter. *The Man Who Fell to Earth*. 1963. Reprint, New York: Avon, 1976.
"Topics of the Times." *New York Times*, January 20, 1941.
Wells, H.G. *Star Begotten: A Biological Fantasia*. 1937. Reprint, ed. John Huntington. Middletown, CT: Wesleyan University Press, 2006.
_____. *The War of the Worlds*. 1898. Eds. David Y. Hughes and Henry M. Geduld. *A Critical Edition of* The War of the Worlds: *H. G. Wells's Scientific Romance*. Bloomington: Indiana University Press, 1993.

BEYOND GOLDILOCKS AND MATTHEW ARNOLD
Interplanetary Triage, Extremophilia, and the Outer Limits of Life in the Inner Solar System

Howard V. Hendrix

Most intensively for the past century and more, Venus and Mars have served as distant funhouse mirrors for our planet Earth. This mirror-staging has helped define our planet's place in what scientists even today refer to as the Goldilocks Zone[1]—namely, while the surface of Venus is presumably too hot and its porridge of atmosphere too thick, and the surface of Mars is presumably too cold and its broth of atmosphere too thin, Earth's soup of atmosphere and surface temperature are "just right" for life as we know it.

The issue of a planet's fitness to sustain life in terms of place and space has long dovetailed with another recurring metaphor of "middle-positioning"—this time a temporal one—for the comparative habitability of the planets of the inner solar system, namely the idea that, like Matthew Arnold's spiritually restless pilgrim in "Stanzas on the Grand Chartreuse," we on Earth are "Wandering between two worlds, one dead,/The other powerless to be born"[2]—the "dead" (or at least "dying") former usually presumed to be Mars, the "powerless to be born" latter, Venus.

There is more to this mirror-staging than meets the eye, however. The past century and more of discussions concerning Earth and its two nearest neighbors suggest that what has been going on in both the scientific and the popular imagination has been a variety of "speculative triage."

Despite the fact that Venus is closer to Earth in terms of size, mass, gravity, and actual physical distance, the funhouse mirror most often gazed into has been Mars, perhaps because the features of its face were, historically, far more discernible to optical telescopes than were those of Venus — at least until the advent of radar and radio telescopy at last pierced the thick Venusian atmosphere.

In contrast, we've been staring at Mars and its seasonally changing features for a long time. From Schiaparelli's telescopic misidentification of Martian features by the Italian word for "channels" and Lowell's mistranslation of that term not as "channels" but as "canals" (and the wonderful Miro/Mondrian Rorschach images Lowell produced in his 1895 book on Mars)[3] — from that time forward Mars has remained the most popular focus of extraterrestrial speculation by human beings. Perhaps (to paraphrase Friedrich Nietzsche) we have stared so long into Mars that Mars now stares back into us.

"Staring" is the correct operative term. Speculative fiction, after all, is fiction about ways of seeing, at least etymologically, given that "speculate" is derived from the Latin *speculari* (meaning "to observe"), itself derived from *specula* ("watchtower" and later "observatory"), in turn derived from *specere*, "to look at, to stare at."

The vast majority, too, of interactions with Mars throughout human history — when not simply speculative — have been optical. In the opening chapters of H.G. Wells's *War of the Worlds*, telescopic observation of Mars is frequently referenced by the novel's narrator and main character, who refers to himself as a "speculative philosopher" and throughout the novel provides us with the human-scale, naked-eye point of view.

In the closing chapters of the novel, although the telescope is mentioned in passing in regard to observations of Venus, the key expansion of optical instrumentality is not the telescope, but the microscope. In Chapter 8, "Dead London," the narrator tells us that the Martians are "— dead! — slain by the putrefactive and disease bacteria against which their systems were unprepared; slain, after all man's devices had failed, by the humblest things that God, in his wisdom, had put upon this Earth" (Wells 136). Our "speculative philosopher" narrator explains this miraculous sparing of humanity thus:

> by virtue of ... natural selection of our kind we have developed resisting power; to no germs do we succumb without a struggle, and to many — those that cause putrefaction in dead matter, for instance — our living frames are altogether immune. But there are no bacteria in Mars, and directly these invaders arrived, directly they drank and fed, our microscopic allies began to work their overthrow [137].

With these passages, Wells first plays that "planets, persons, pathogens" chord that continues to echo down the years to our own time. Since Wells' foundational text chronicles a *war* of the worlds, rife with images of invasion (and of Earth's unconscious counterinvasion of the Martians through bacterial infection), it's not a giant leap to realize that Wells' underlying metaphor of the relationship among the inner planets is one derived from the battlefield, namely the idea of triage.

Although "triage" derives from the French *trier*, "to sort," there is also something inherently "tri-" and "triple" about it, since triage is fundamentally a sorting into three, a system designed to produce greatest benefit from limited resources for battlefield casualties by *denying* treatment both to those who have no chance of sustaining life and to those who will be able to sustain life without treatment, and by *providing* available treatment to those who may be able to sustain life if given that treatment. The speculative triage scenario revises the Goldilocks paradigm's middle position. Too-hot Venus remains on the outs as the world with no chance of sustaining life, but just-right Earth, as the world with no need for special treatment in order to sustain life, also falls outside the treatment category. Too-cold Mars is now the triage-middle world, the one that will most benefit from special treatment in order to sustain life. In Wells's novel, this special treatment paradoxically involves invading a "healthier" world, a concept to which we will return later.

Interplanetary triage continues with Olaf Stapledon's *Last and First Men* (1930). His Martians are as rapacious and (initially) very nearly as amoral as Wells's "demonic other" Martians. They are also, if anything, even more alien. Unlike Wells's Mars, which has no microbes in it, Stapledon's Martians *are* microbes: "ultra-microscopic subvital members," "smaller than terrestrial bacteria or even terrestrial viruses" which form a macroscopic "cloudlet, a group of free-moving members dominated by a 'group mind'" (136). Stapledon's Martians are pathogens which have achieved a sort of personhood.

In contrast, too, to the brief encounter Wells chronicles in *War of the Worlds*, in Stapledon's *Last and First Men* the war between humans and Martians goes on for millennia upon millennia. Only gradually does each of the opponents begin to grasp the true alienness of the other. In Stapledon's retelling the long-stalemated doomsday war between Earth and Mars comes to its close when a supergerm is purposely and quite consciously released by humans, the Martian colony on Earth and then all the Martians back home are destroyed by it, and by it humanity itself is driven to the brink of extinction—not quite over the edge, but so far as to descend into an epochal dark age. After vast stretches of time, the remnant "Martian subvital units that had been disseminated by the slaughter of the Martian colony and had then tormented men and animals with pulmonary diseases" (167) gradually coevolve with terrestrial life-forms, becoming symbionts—literal "microscopic allies," to use Wells's earlier term—which in Stapledon's novel eventually give to humans some of the same quasi-telepathic abilities the Martian cloudlets once possessed.

Both Wells's and Stapledon's visions of the Martians make use of the "Mars as dying planet" hypothesis put forward most persuasively by Percival Lowell. Their Martians are embodiments of their planet's supposed senescent condition—life-forms originating on an aged, dying, depleted planet, the world most in need of another world's resources in order to survive and thrive. In *War of the Worlds*, Wells makes the point that it is Earth's *putrefactive* bacteria that destroy the invaders: The Martians are dead meat because, as far as Earth's bacteria are concerned, the Martians were *already* dead meat. Perhaps the reason the Martians succeed in their invasion of Venus is that a planet where life is "powerless to be born" is even more dead than a planet that has died or is dying. Mars here is the world in the triage middle, neither too far gone to benefit (like Venus) nor so able to continue sustaining life as to need no help (like Earth).

On this same life-and-death spectrum, Stapledon consistently describes his Martians as based on "subvital units" out of whose interactions life-like characteristics seem to arise as a sort of emergent property—which makes it difficult to say whether they were ever actually living, in a terrestrial sense, at all. And although human beings in both novels may be said to "triumph" over the Martians (to whatever degree or for however long or brief a time), their encounter with the Martians has Martianized humanity: literally and biologically in the sense of the human–Martian "hybrid" in Stapledon's novel, figuratively in the way in which Wells' narrator sees "the busy multitudes in Fleet Street and the Strand" as "but the ghosts of the past ... phantasms in a dead city, the mockery of life in a galvanised body" (Wells 147)—a humanity like the benumbed speaker in Emily Dickinson's "After great pain a formal feeling comes —," a humanity in a sense already dead like the already dead Martians, only not yet awake to the fact of their decease.

Given such an interlinked Borromean knot between and among planets, persons, and pathogens—a complex so centered around notions of the Other, with Death in the role of ultimate Other—it's not surprising that the medical and microbial readily scale up to the political and personal, particularly in the context of invader vs. invaded, colonizer vs. colonized. Just as a triage-middle Mars can be seen as requiring the "special treatment" of invading a "healthier" world in order to sustain life, in the long grow-or-die emergency that is human history, empires too can be seen as requiring the "special treatment" of invading and colonizing "underexploited" lands in order to sustain their ways of life.

The fundamental "What if?" of Wells's *War of the Worlds* can be rephrased as "What if, in the last decade of the nineteenth century, technologically advanced invaders from space landed in Britain and did to the British what the nineteenth-century British Empire did to peoples throughout Asia and Africa?" Many critics have commented on *War of the Worlds'* "brutal allegory of colonialist exploitation," confronting its readers with "the implications of racism and colonialism" (see Markley's chapter "Wells's *War of the Worlds*: Apocalyptic Disintegration" in his *Dying Planet*, particularly pages 122–125). Wells also suggests in the text that, having suffered what its own colonies have suffered—invasion, conquest, colonization, expanding genocide—Britain will, as a result of this extraterrestrially forced empathy, cease to be a colonial power, and colonialism itself will cease to exist. That, at least, is the hope held out by the narrator's statement that "the invasion from Mars ... has done much to promote the concept of the commonweal of mankind" (Wells 144).

Plausible as all that may or may not be, Wells undeniably gives his British readers an out by granting a distinctly counterhistorical reprieve: Intriguingly, he inverts the history of microbes and conquest. The more technologically advanced Martian invaders are defeated by the microbes of the less technologically advanced invaded. This is in marked contrast to the history seen on Earth, in which the microbes accompanying the more technologically advanced European colonial powers helped them conquer and subdue the less technologically advanced peoples they invaded, rather than the opposite (as Jared Diamond's *Guns, Germs, and Steel* makes amply clear). Stapledon's Martians too are ultimately defeated by a microbe associated with the invaded, but in *Last and First Men* there is nothing accidental or unintentional about the release; the "murder" of the Martians is thoroughly premeditated.

In *The Martian Chronicles*, written a generation after Stapledon's *Last and First*

Men, Ray Bradbury gives a new twist to this knot of planets, persons (both human and Martian), and pathogens. In *The Martian Chronicles* the world invaded is Mars rather than Earth, and the microbes accompanying the human invaders are the source of Martian collapse. In the pivotal chapter entitled "June 2001:— And the Moon Be Still as Bright," we learn that what killed the Martians was

> chicken pox. It did things to the Martians it never did to Earth Men. Their metabolism reacted differently, I suppose. Burnt them black and dried them out to brittle flakes. But it's chicken pox, nevertheless. So York and Captain Williams and Captain Black must have got through to Mars, all three expeditions. God knows what happened to them. But at least we know what they unintentionally did to the Martians. (50)

In highlighting the history of unintentional decimation, Bradbury interestingly echoes Wells: In *The Martian Chronicles*, Earth's microbes are more alien to the Martians than the Earth Men themselves are, for at least the Martians can communicate with humans, even if the results of such human–Martian interaction during the first three failed expeditions (York's, Williams', and Black's) led to the death of the humans, often through their going strangely native. Such abortive attempts at settlement remind the reader of the history of European colonial powers in the New World — powers even whose failed colonies succeeded in decimating the indigenous populations with smallpox and other diseases which accompanied the colonists.

In the same "June 2001" chapter, Bradbury makes all the clearer that his book functions as an allegory or displaced retelling of the history of the European colonization of the New World. One of the explorers, Cheroke, says "I've got some Cherokee blood in me. My grandfather told me lots of things about Oklahoma Territory. If there's a Martian around, I'm all for him" (59). Here Bradbury specifically links the fate of the inhabitants of the "Red Planet" to the fate of the "red man" in the Americas.

Yet even here allegories and histories of colonizer and colonized don't remain that simple. After the gone-native Spender, using a Martian "bee-gun," begins killing off his fellow crewmembers — essentially as punishment for their insensitivity to the Martians' plight — he asks Cheroke to join him in his crusade against his fellow colonists. When Cheroke refuses, Spender shoots him (60). The gone-native would-be colonial, who feels compelled to protect by any means necessary any remaining Martians from a final genocide, ends up killing, with that Martian weapon, a descendant of a Native American people already genocidally colonized.

The implications and ramifications of such a scene are well suited to a postcolonial interpretation, but for the sake of brevity I will focus here only on how it is that Spender goes native. He claims to have unintentionally "found a Martian." As a result of Spender's "learning how to read the ancient books and looking at their old art forms," a Martian begins appearing to him sporadically, "until on the day I learned how to decipher the Martian language" the Martian appears before him permanently and takes possession of his soul — which ultimately leads Spender to begin killing his fellow explorers (59).

In contrast to Wells's "demonic other" Martians, the Martians of *Chronicles* are the "exotic other," fascinating and attractive. Yet, like Wells, Bradbury also intriguingly inverts the larger colonial history. As Wells's Martian invaders of Earth are overcome by the microbes of our invaded planet, so too are Bradbury's human invaders of Mars initially (and also perhaps ultimately) overcome by the memes, the mind-viruses, the dreams, thoughts, arts and culture of the Martians. Spender is seduced into becoming a Martian, into going native in the fullest sense, by his interaction with Martian cultural artifacts. It is precisely on the day he learns to decipher the Martian language that a Martian captures his soul.

Spender's "going native," for all that it is mediated by language, is individualistic and unintentional. The use of language in the colonial situation, however, has been more often collective and intentionally coercive. In his book *Decolonizing the Mind*, Kenyan novelist, playwright, and critic Ngugi wa Thiong'o writes of the European colonization of Africa: "In my view language was the most important vehicle through which [colonial] power fascinated and held the soul prisoner. The bullet was the means of physical subjugation. Language was the means of spiritual subjugation" (9). *The Martian Chronicles* inversion is that Spender the invader, mind-virused (co-opted, subjugated) by the conquered Martian culture, follows Martian orders to kill his fellow Earth Men — rather than the more historically common incidence of, say, Ibo officials in Nigeria following British orders to kill their fellow Ibo, as in Chinua Achebe's 1959 novel *Things Fall Apart*).

I will not here go so far as to say with William S. Burroughs that language is a virus from outer space, but I think there is a good deal of truth to Ngugi wa Thiong'o's statement about his own colonized African experience that "language and literature were taking us further and further from ourselves to other selves, from our world to other worlds" (14). In the colonial and postcolonial African context, he is indeed saying this "like it's a bad thing," yet I don't think it must necessarily be so.

In *The Martian Chronicles*, Bradbury has moved the just-right Earth of the Goldilocks middle back to the middle of the triage scenario as well. Life on Earth has become as unsustainable as empire itself — as a result of the threat and eventually the reality of Earth's self-destruction through nuclear immolation. Paradoxically, by the end of the book, Mars becomes more capable of sustaining human life than Earth itself.

In *The Martian Chronicles*, the human colonizers do indeed move farther and farther away from their old selves and old world — until they become the other, become the Martians, most powerfully through imagination. This process is seen not only in "June 2001" and in the final chapter, "October 2026: The Million Year Picnic," but also in "August 2002: Night Meeting," which raises questions about the entire temporal supposition that Mars and its inhabitants are the dying past while Earth and its inhabitants are the growing future. (This theme returns at the end of the text with "Million Year Picnic" and its story of refugees beginning their lives over on Mars after humanity has largely obliterated itself via nuclear war on Earth.)

In "Night Meeting," Earth Man Tomas Gomez tells Martian Muhe Ca, "Mister, you're invaded, only you don't know it. You must have escaped" (83). When Tomas says, "If I am real, you must be dead" (82) and "But the ruins prove it! They prove I am the Future, I am alive, you are dead!" (85) he is essentially telling Muhe Ca, "Mister, you're dead, only you don't know it yet." Like the Martians in Wells' *War of the Worlds* from the standpoint of the putrefactive bacteria. Like the cloudlet Martians in Stapledon's *Last and First Men* from the point of view of terrestrial biology. In the triage scenario, the dead and the past are both beyond help, the future and the not-yet-born can take care of themselves, and none of them really merit worrying about.

And yet in all these cases the past turns out to be not quite dead, and the future is not quite powerless to be born. Even if Gomez and Ca literally cannot physically touch each other because they do not exist in a shared physical present between past and future, the living are nonetheless touched by the dead, whether it be Wells's Fleet Street Londoners, Stapledon's Second and Third Men, or the human family who recognize themselves as future Martians in "Million Year Picnic."

Although much could be said here about the thinking Martians who project their thoughts into fetal humans in order to colonize humanity in Wells's *Star-Begotten*, or about C.S. Lewis' Perelandra trilogy as a religious variant on the Goldilocks model, I am constrained to move on to the late 1970s—a generation after *The Martian Chronicles* first appears and after the *Viking* Landers have done their initial life-search on Mars. The emphasis in Mars fact and fiction shifts away from the middling position for Earth and its inhabitants and toward the extremes: to remaking human beings from the cellular level on up to fit the Martian environment (as in Fredrik Pohl's *Man Plus*), or radically remaking Mars to fit human beings (a process in which the microbial realm again figures prominently, as in Robinson's *Red Mars*, *Green Mars*, and *Blue Mars*).

With this shift—and an accompanying movement of Mars speculation beyond the realm of science fiction entertainments and into organized Mars boosterist groups and scientific program planning—the planets/persons/pathogens complex alters slightly. What were once focused on as pathogens are now more often seen simply as microbes and are more likely to signify the presence of life than the threat of death. Mars itself is now presumed to be far more extreme a world than in the days of Lowell, but we've also discovered that life abides in the most extreme conditions of Earth.

What has altered more substantially is the entire underlayment of speculative triage. Triage is always about the allocation of scarce resources, and traditional triage emphasizes the allocation of resources to that group found between the extremes of those who have no chance of survival given available treatment resources and those who can survive without need of those resources. However, the fears of the 1980s that global war and nuclear winter might push our climate toward the extreme of Mars, or more recently that human-assisted global warming will push Earth's climate toward the extreme of Venus, have resulted in a shift toward a "reverse triage" paradigm. In this variant—also first developed for the battlefield—the more superficially injured are treated first so they might return to battle, while the more severely wounded are abandoned. In the reverse triage model, the emphasis is on allocating resources first and foremost to preserve Earth's fitness to sustain life "as we know it," even though it is already more fit to sustain life than Mars is, and far more fit than Venus.

Over the last thirty years, this shift has meant that the emphasis has been more on how Mars and Venus might provide models of Earth's past and future, rather than the importance of Martian or Venusian environments in themselves. During this time, human and microbial thrusts concerning Mars missions have expanded with the search for Mars analogs. According to the Mars Society document *Expedition Two: An Interdisciplinary Mars Analog Research Expedition to the Arkaroola Region, South Australia*, a human-scale Mars analog is defined as "an environment or situation on Earth with characteristics, in nature or by simulation, for which there are, or could be, analogous characteristics on Mars" including "both the physical setting of Mars, as well as design considerations for technological challenges and scenarios for human activity" (4). Such Mars analog expeditions — in many ways simulated colonies which bring Mars to Earth rather than Earth to Mars — have already taken place at the Mars Desert Research Station (MDRS) in Utah, Flashline Station (FMARS) at Houghton Crater in the Canadian Arctic, MARS-OZ at Arkaroola in Australia, and ongoing efforts toward Euro-MARS Iceland (4, 15).

Although Mars analog expeditions have previously placed a strong emphasis on geological conditions and human colony factors, a growing rationale (again according to NASA and Mars Society documents) for microbial observatories in Mars analog locations is again essentially about researching Earth first, above and beyond colony simulation: namely, that "current ecosystems can provide models for possible extinct or extant Martian ecosystems. Since development of these methodologies on Mars will not be easy, it is best to develop methodologies for life detection here on Earth, prior to our exploration of Mars" (15).

Life used to be seen as constrained to temperatures ranging essentially from the freezing point of water to its boiling point, on a wet substrate at one atmosphere pressure, lit by sunlight whose harmful hard radiation has been screened out by a moderately thick atmosphere. Not anymore. The age of precolonial simulation is the age of extremophiles: acidiphiles, alkaliphiles, halophiles, piezophiles, psychrophiles, thermophiles, xerophiles; anaerobes, endoliths, and oligotrophs — all of which live on Earth in conditions humans consider, in one way or another, "extreme." This, too, is a change in the scientific literature, at least rhetorically: Environments once labeled with the demonic-othering of "hostile" are instead now "extreme" or "exotic."

By setting up microbial observatories in environments which may be in at least one way or another like certain environments on Mars, we have broadened our understanding of the diversity of life on Earth, ultimately serving to make Mars and Earth look more like each other at their extremes than previously assumed.

Yet extremity is relative. Many approaches for adapting to extreme environments are known from Earth, including adaptations to some of the same stresses that occur on Mars (cold temperatures, low pressure, lack of liquid water, and an intense radiation environment). Despite our "Fix Earth First!" rationalizations for researching such environments, however, the combination of stresses on Mars remains unique and is not exhibited in any single Earth environment.

For all the mirroring, Mars remains Mars and Earth remains Earth. If, like Gomez and Ca in "Night Meeting," life on Mars and life on Earth cannot occupy the same present, then we can always look for Martian life in the past (as in the controversy over whether or not the Martian meteorite found in the Allen Hills in Antarctica contains fossilized Martian life or not, and whether or not life on Earth actually originated on Mars). Or look for it in the future: if microbial life does not exist on Mars (à la Wells), it may be necessary for us to invent it — perhaps a fitting challenge to the nascent discipline of synthetic biology.

We may say with Ngugi wa Thiong'o that language and literature take us further and further from ourselves to other selves, from our world to other worlds. The history of speculative literature and practical exploration concerning Mars, however, shows that we are also brought back by such imaginings, returned to ourselves and our world, in the sense that Eliot suggests in the last stanza of "Little Gidding" in *Four Quartets*: "We shall not cease from exploration/And the end of all our exploring/Will be to arrive where we started/And know the place for the first time."

I will end with a story. For a few hours on May 25, 2008, I left the Bay Area Science Fiction Convention (BayCon), where I was a guest, and made the short trip to NASA Ames Research Center in Mountain View, California, to join the scientists and enthusiasts who had gathered there in anticipation of the entry, descent, and landing of the *Mars Phoenix Lander* on the surface of the Red Planet. There we endured the curiously post–Einsteinian suspense of anticipating the past: waiting to discover the success or failure of a landing that had in fact already happened fifteen minutes earlier on the surface of a different planet — not to mention what the scientists themselves called the "seven minutes of screaming terror," that period during the descent which was both the most fraught with potential pitfalls and unavoidably radio-silent.

In such a spacetime-twisted, discontinuous, disaster-in-a-funhouse world, one can dream in a Night Meeting that the extreme and exotic environments we continue to search out will not only be for the preservation of this world, but also in preparation for a fuller, on-site understanding of another. One can dream that the other — "Ah yes, that other over there, unreal, a ghostly prism flashing the accumulated light of distant worlds" (Bradbury 82–83) — will also be found in ourselves. One can dream of a time when the knot of planets, persons, and plagues that has been the imagined triage relationship between Mars, Venus, and Earth at last transitions to a time like that which the narrator describes in the last pages of *War of the Worlds*: "Dim and wonderful is the vision I have conjured up in my mind of life spreading slowly from this little seed bed of the solar system throughout the inanimate vastness of sidereal space. But that is a remote dream" (146). Yet one can dream, one *must* dream — for, as the philosopher Gaston Bachelard put it in his *Poetics of Space*, "If we cannot imagine, we cannot foresee" (xxxiv).

Notes

1. See, for instance, "The Goldilocks Zone," from the April 28, 2007, issue of *Astrobiology Magazine*, based on a European Space Agency (ESA) press release. Available online at http://www.astrobio.net/news/article2314.html.

2. "Stanzas from the Grand Chartreuse," lines 85–86. First published in *Fraser's Magazine* (London), 1855.

3. Illustrations of Martian surface features from Lowell's text are readily available online, including at http://www.bibliomania.com/2/1/69/116/frameset.html.

Works Cited

Bachelard, Gaston. *The Poetics of Space*, trans. Maria Jolas, new intro. and foreword, John R. Stilgoe. Boston: Beacon Press, 1969, new foreword, 1994.

Bradbury, Ray. *The Martian Chronicles*. 1950. Reprint, New York: Buccaneer Books, 1977.

Markley, Robert. *Dying Planet: Mars in Science and the Imagination*. Durham, NC: Duke University Press, 2005.

Stapledon, Olaf. *Last and First Men: A Story of the Near and Far Future*. 1930. Reprint, Los Angeles: Jeremy P. Tarcher, 1988.

Wa Thiong'o, Ngugi. *Decolonising the Mind: The Politics of Language in African Literature*. London: Heinemann, 1986.

Wells, H.G. *The War of the Worlds*. 1898. Reprint, London: Dover, 1997.

Appendix 1

To Write the Dream in the Center of Science: Mars and the Science Fiction Heritage

A Dialogue Between Ray Bradbury and Frederik Pohl (George Slusser, Moderator) (May 2008)

George Slusser (GS): A couple of events are very, very memorable to me. One was when I was in an old-fashioned public library such as Ray Bradbury has described. I guess I was about nine years old. I opened *The Martian Chronicles*. I had no idea what this was, but the title sounded interesting. The first chapter, the little chapter, bears the title "Rocket Summer." And there I read the sentence: "Rockets make climates." And I said to myself: What the hell is this? Nine years old, and you come into contact with an idea like that. I'm a native Californian. I had never seen Ohio. I didn't know what snow was. But I imagined a place of dead white winter, where suddenly and briefly, it was warm and like California. This single sentence may have changed the way I saw the future and its possibilities. It certainly launched me into a passion for SF; with one stroke, it threw all the parameters of my existence off into a different realm.

Another event, later on, much later on, was when I read Fred Pohl's novel *Man Plus*. This of course was a very different age and time. But from that novel I realized how extremely difficult it would be actually to go to Mars, and began to wonder if it would be worth all the pain and expense to make the trip. And the trip to what? Ray Bradbury opened my mind to sense of wonder. Fred Pohl made me open my eyes. This was the end of the Golden Age, and the beginning of an age of realistic limits. Do we want to be "man plus"? Mars is not just a dream or a metaphor, it's a place and it's cold as hell. Rocket winter.

Now I am honored to sit next to these two great writers, so important to my intellectual development. I would like to establish a conversation between. I want to ask them a question and have them each answer it and maybe we can start talking and I will simply back away from it.

Frederik Pohl (FP): George, before you do that, can I ask Ray a question?

GS: You certainly can. I'll get out of this.

FP: Ray, I was interested to hear you tell the audience that modern science fiction's all trash. [Laughter]

Ray Bradbury (RB): Not completely...

FP: Ah. But there are, there were some science fiction writers here this weekend. There was Greg Bear, Greg Benford, David Brin, Kim Stanley Robinson, and between them they've written a lot of books. Which of those books have you read? [Laughter]

RB: Well, I have personal knowledge of Greg Bear's books, because I've known him since he was 16 years of age. And I encouraged him, because he was an artist. I encouraged him to become a writer and I believe he's doing good books, but the market was mainly ruined because of bad movies.

FP: I think you're right about that.

RB: It's not the fiction so much as what's being done in the studios. The studios do not give a damn for us. The Spider Man series made a billion dollars.

FP: Garbage.

RB: You can't compete with that.

FP: Ted Sturgeon used to have a saying. He said that 90 percent of science fiction is crud. But then, 90 percent of everything is crud.

RB: That's true.

FP: (to GS): OK, you can have the floor back now.

GS: I will ask a question. Both of you gentlemen have written definitive novels about Mars. And I would like to know today, given what we know about Mars and what we know about the way things have evolved in our culture, would you write a Mars novel today? Either of you.

FP: You mean a realistic one?

GS: Any novel. Would you choose Mars as your subject or would you choose something else?

FP: There are things I would like to say about Mars. But the thing that Ray wrote about, which was the *discovery* of Mars, it's past the time for that.... You can't talk about somebody landing on Mars for the first time now, because it's happened. It's old news. And we know too much about Mars to write about Barsoom and Dejah Thoris, and the six-legged, six-limbed great white apes and all that. So I wouldn't write that kind of a story. But I would set some stories on Mars, and I'd try to make them realistic, but then you're very limited, for as everybody has said, Mars is not friendly to life, really. So you can't really have people walking around there as they do in Riverside, California. So I would hesitate.

GS: Mr. Bradbury, would you write a Mars novel today?

RB: Oh, God!

FP: Done that!

GS: Say you had never written one and you were faced with the possibility of writing one today, would you accept to do it?

RB: I'm writing an article for the *National Geographic*. It will be out in the next few months. It's partly an article and it's partly *The Martian Chronicles*. Which is to say, would I write some more about Mars? You're damn right I would! The things I've said today will be in that article. I will be writing more articles. I'm doing a novel on the religions of the future. It will deal with life on Mars. So I'm already doing it. I don't know where the hell it's going to go. I'm not in control. So, we've got a long way to go. We're thirty years behind. We should never have left the Moon, we should have made it our base for Mars, but we didn't do it. Now we've got to reestablish ourselves. So if I were a young man today, I'd be writing my version of *The Martian Chronicles*, to get us back to the Moon and to get us to Mars. So we've got a big job. And the answer is "yes." My younger self says "yes." And I'm going to finish my book on the religion of space in the next three years. It will start on Mars and it will end on Alpha Centauri.

GS: Wonderful. Fred, you have a comment.

FP: I'm glad to hear you tell us that we're going to the Moon and then to Mars, because I'd begun to have my doubts. This magazine, *The Planetary Report*, which is published by the Planetary Society, has a long article which I just read, saying that the present American plan, the present administration's plan, is to build a humongous moon station and use that as the getting-off place to go to Mars. But the moon station is going to cost so much there will be no money left to go to Mars. The other thing that I've heard and I'm not sure if it's permanently, eternally true, is that human beings are not likely to get to Mars because the risks of being killed by a solar flare or cosmic rays is so great that nobody can survive it unless they find some way of shielding themselves from the radiation. But I wish it were true. I would love to go to Mars. I would love even to see some other person go to Mars if I can't do it myself. I just have a fear that it's not likely to happen within the next few centuries.

GS: I have another question that interests me. The term "science fiction" resulted from a contest in *Amazing Stories* in 1926 to name the kind of fiction Hugo Gernsback was publishing. The genre has changed a lot since. How would you describe what science fiction has become today? Can I start with Ray?

RB: To write about a dream in the center of a science. Science fiction starts with an egg at the center, and compels scientists to write more about science, so we're helping each other. The scientists like to be with me because I talk dreams. I dare to talk dreams. They can do the implementing; I can't do that. Someone like Asimov knew more about science than all the other writers. He was fantastic. We both started at the same time. So he was a companion of mine. We knocked elbows. But I couldn't write his fiction and he couldn't write mine. So we needed each other. It's a combination. The dreaming of doing a thing that began outside a cave a million years ago. We are destined to do this. We're not in control. It's part of our ecosystem. I can't explain it. But it's there. If I can begin that way and we are here today to applaud the way I've gone, I wasn't sure I was going to get here, the science fictions, the worlds that we indeed represent, the names you mentioned here today. I would not dream of offending them. They are fine writers. But the other writers are so influenced by movies, they're writing the wrong science fiction. They're not being given a chance to write anything except Spider-Man and Superman. But that isn't science fiction. That is the real trash. Giving you an example of what's happening with me and Hollywood right now. Universal Studios optioned *The Martian Chronicles* ten years ago. There are twenty-six scripts. It's all trash. It's all junk. It's written by rock musicians. [Laughter] And the film hasn't been made. I keep saying that for God's sake I'm 87 years old. I'm not going to be around if you make the goddamn film. *Fahrenheit 451* was optioned ten years ago by Mel Gibson, and there are twenty scripts on that. You see, the filmmakers have no respect for quality. And the young writers coming up through science fiction. They try to do things for Hollywood. That's the wrong direction. I'm sure that Brin and Benford and Greg Bear are trying to carry on doing what they've done; it's gonna be beautiful. But I warn people to be careful what they look at as science fiction. You've got to find the real dreamers — and Greg Bear is one of them. You know he's going to do it. Who knows who it's going to be when the next writers grow up. Because right now our political system, both the Democrats and the Republicans, they don't give a damn. We have had to go through a hundred days of speeches and there's a hundred fifty more days of hot air. Right? And both parties are bankrupt. The Democrats and the Republicans are both bankrupt. So, there are no people there who care about space travel. Thirty years ago there were several senators who had a real love of science fiction. Beautiful. But there are no senators like that today. You've got to find out who to elect and send them to Washington to represent us and get us to the planet Mars.

GS: Good luck.

FP: You mentioned, Ray, the difficulties you've had with people optioning your books and getting manuscripts. The book of mine that's been mentioned here is *Man Plus*, it's been under continuous option or ownership to Fox Productions for about twenty-five years, or a little more, ever since shortly after it came out. For years they would pay me option money, an option payment every year, which I began looking on as my annuity. [Laughter.] Because every year it got a little bigger. And then they double-crossed me; they went and bought it outright. But they've never made it and it doesn't seem as though they ever will. I am told they spent over a million dollars on scripts for it, without getting a good one. They've asked everybody they could think of how to make this film, with one exception. There's one person they've never asked. Me. [Laughter.] Anyway, I think your original question was "What is science fiction?" And the answer is, science fiction is the name we give to that kind of stuff we like to read. It's a name. It's not a definition. It's not a description. My name is Fred. It doesn't describe me, it's just what people call me. Science fiction is what we call that part of the bookstore or that part of the library. What it's about is looking at the world as though something had changed and what would it be like if these things had changed. Either you're going into the future, or the Martians have landed, or there's a plague that wipes out ninety percent of the earth. Whatever it is, we try to examine what would happen in our world if this became true. The writer is allowed to tell one big lie. But after that it all has to be developed logically. And that's science fiction.

GS: One question: Do you have a science fiction film that you like? Either of you.

FP: Yeah, I think you mentioned it. It's called "Things to Come" and it was made sixty years ago. [To RB:] Is that the one you're talking about? Yeah, "Things to Come." I've seen it 38 times. I saw it again a couple of months ago. It was on the Ted Turner classics television station. It's really a crude movie, and it doesn't have a good story, but it has a lot of wonderful ideas in it. And that's what most science fiction films lack. They have no ideas. What they have is people killing each other with ray-guns.

GS: Does anyone in the audience have a question for either Ray Bradbury or Fred Pohl?

FP: May I first make a comment? As you can see, Ray and I sometimes disagree on things but there is one thing that I know we agree on. And that is that science fiction produced one thing that was of immense importance to both of us. And that was science fiction fandom. We learned to be science fiction fans and we learned how to write by trying to write for fanzines, and by talking to other fans and learning, and that has been the way that the best science fiction writers have come along. And I say, "God bless the science fiction fans!" [applause]

GS: Please give a big hand for these two pillars of the science fiction field. [standing ovation]

FP: We appreciate your standing up, because Ray and I can't.

Appendix 2

THE EXTREME EDGE OF MARS TODAY
A Panel Discussion with David Hartwell, Geoffrey Landis, Larry Niven, and Mary Turzillo, Moderator (May 2008)

Mary Turzillo (Moderator): Right! I'm a founding member of the Mars Society and my novel and my short stories about Mars are extrapolated from hopefully realistic ideas, generated by Robert Zubrin and also my own husband, Geoffrey Landis, about colonizing Mars, and I would like to introduce these people who are all among my heroes, definitely my heroes.

You all know Larry Niven. And the two words that occur to one — four words for Larry Niven — are big ideas, living legend. An early Mars story of his was "At the Bottom of a Hole." He also has a Mars novel, *Rainbow Mars*. I can't possibly tell you everything about this incredible man, but he is a Caltech alum, you know he won a whole bunch of Hugos, I forgot how many, you know a lot [Larry raises five fingers] thank you, and among his other hobbies are "saving civilization and making money, moving mankind into space by any means, particularly by making space endeavors attractive to commercial interests, if that's something of interest to you." It's interesting to me. And he's done a lot to promote this. He's been [a] member of committees that were advising Ronald Reagan for national space policy, and he's had some effect on the space program. The SDI Space Initiative or Star Wars was drafted at his house in Tarzana. I'm just hitting a couple of things here. On the stands right now are *Scatterbrain*, a retrospective anthology; *Burning Tower* with Jerry Pournelle; *Creation Myth* with Brenda Cooper; *Ringworld's Children*— Doesn't that sound like a treat?—*Draco Tavern* (which I'm probably mispronouncing), twenty-one ultra-short stories; and there's just so much more.

Next, Geoffrey Landis, who I met in 1985 when we were both at the Clarion Science Fiction Writing Workshop. I liked his ideas so much that I thought I would marry him, so I did. You know, I do know about him, but I wanted to know what he was willing to share. Dr. Geoffrey Landis knows Mars both as a scientist and as a science fiction writer. His novel, *Mars Crossing*, which is available here [at the April 2008 Eaton

Conference at the University of California, Riverside], has been called the most accurate novel about Mars ever written. His short story, "Falling onto Mars," won the Hugo award for best short story in 2003. He has two Hugos and one Nebula. He is also a member of the *Sojourner* rover team, on the Mars-Pathfinder mission, and he's currently a member of the Mars exploration rover science team, driving two rovers, *Spirit* and *Opportunity*, around the surface of Mars.

Then we have David Hartwell, who has a doctorate in comparative literature from Columbia University. He's won the Hugo for best editor and he's been a Hugo nominee 33 times to date. He's the publisher of the *New York Review of Science Fiction*, which you should all subscribe to ... it's a wonderful publication.

David Hartwell: I have flyers.

Moderator: Well, I've got one up here, but there's more ... and he's a senior editor at Tor/Forge and, what else did I want to say? Sometimes people are too modest to say the things that people really ought to know. I would say of him that he has an encyclopedic mind, perhaps the most encyclopedic mind in science fiction, but it's an encyclopedia which interacts and generates shiny new ideas. He's the co-editor of the *Hard Science Fiction Renaissance* with Kathryn Cramer.

Okay, I have a whole bunch of questions with which I'm going to grill these people and I'm going to start with this: What is your Mars like? Has it changed with knowledge of the new Mars?

Larry Niven: Yes!

Geoffrey Landis: Well, that was a succinct answer.

Niven: I started early. I started writing Mars stories in the '60s, when the results were just coming in from probes. Somebody was talking about sand dunes on Mars, so I wrote a short story involving that. And then pictures came back and the planet was covered with craters, just like the Moon. That was startling, and I joked to be the first to write a story called Mars with craters. More data kept flowing in and at that time the computers hadn't spread like a great plague or a great symbiot, for that matter. The computers hadn't spread, the Internet wasn't there, you had to fight to get information. I could be the first with every new discovery about Mars, for about six years. And then it kind of ran out and I started looking elsewhere.

That was then. The most recent thing I've done about Mars was *Rainbow Mars*, and you have to get this in perspective. Hanville Svetz is a time traveler, and time travel is fantasy and he doesn't know it. So the Mars he's finding contains everybody's Martians — everybody's except Heinlein's, who were too powerful; I couldn't handle those. Why did I do fantasy? Because the real Mars is available to everybody; I can't be the first at anything. I've tried it with other things such as the frozen Earth and found I got beat into print. That is how Mars has affected me, accepting the fact that I can keep updating it as information flows in, and it's still part of known space, part of the solar system I've been writing about for forty years.

Landis: Okay, well, my fiction has been evolving as we learn more about Mars, and I'm not even sure sometimes what to think about the Mars that we're now discovering. I guess my first story that really featured Mars was a story called "Ecopoiesis," which is in *Impact Parameter* which is now sold out in the dealers room, but I'm sure you can find it online. And that had a sort of odd genesis. I was at a Case for Mars conference in, I think the mid-1980s, and Carl Sagan had a paper about terraforming Mars, and there was a lot of discussion about terraforming Mars. There's been a lot of very inaccurate science fiction about terraforming, most of which proposes that it's just vastly simpler than it really is.

It's a difficult problem, but the idea had been sort of floating around that what would be interesting to do was not to necessarily terraform Mars in the sense of "terra" meaning "make Mars like Earth," but just to set an ecology on Mars, not necessarily an Earth-like ecology but make it a life-filled planet, and the word for that is "ecopoiesis"—to initiate an ecosystem. And it would be an anaerobic ecosystem because Mars has no oxygen, and the interesting thing of course is that if you just want to warm up Mars you don't want to give it an oxygen atmosphere, that would be the worst possible thing to do because the carbon dioxide atmosphere is the only thing that keeps it as warm as it is, and its average is -40°. So you don't want to get rid of the carbon dioxide; you want more carbon dioxide. You want a thick carbon dioxide atmosphere. So ecopoiesis and warming up and giving an ecosystem to Mars would in fact mean adding more carbon dioxide and making an anaerobic ecology. And Sagan was sort of a little bit sarcastic about this possibility and said, "Well, let's see, let me see if I get this straight—basically, you're proposing converting Mars into a sewer. Anaerobic bacteria! Yay, sewer bacteria!" So, the sort-of working title of *Brown Mars* was going through my head. So this was a story about Mars that had been not terraformed, so it wasn't Earth-like, but yet it had an ecosystem, an anaerobic ecosystem that, as our astronauts going to it many years after that ecopoiesis event, take off their space suits and say, "Oh my God, Mars smells like shit." So that was my first Mars story. It did incorporate one thing from all of the spacecraft results: I was looking at them saying, "My God, Mars is a sulfur-rich planet." Mars *is* a very sulfur-rich planet, so I put sulfur in as a major plot element in the story and am sort of pleased every time we send another mission to Mars and the chemical analyses we're getting over and over [with] this emphasis that Mars is sulfur rich.

So I said, yay—I got that one! The salt that's left behind from the putative vanished oceans or perhaps the very briny lakes, the salt that we see in the salt stones and silt stones on Mars is a very sulfur-rich salt, it's calcium, magnesium, iron sulfates, so it's not sodium chloride, which would be most of what would be left on Earth if we evaporated our oceans. So, you know, it really is a very rich planet in sulfates, and that probably comes from its long history of volcanism, that Mars does not have global plate tectonics that can bury the sulfates and get them out of the shallow crust layer. I guess sulfates were a little bit also a plot point when I finally wrote a novel about Mars—*Mars Crossing*—where I tried to write a novel about Mars that was accurate. Seems [that in] too many Mars novels ... you get to the end [and] you discover either (a) the ruins of an ancient Martian civilization or (b) artifacts from aliens that have come to Mars. I said, you know, the real Mars is also really interesting; there's a lot of interesting stuff!

So I tried to put as much of Mars as we knew at the time into *Mars Crossing*, showing that it's not Earth but it is a fascinating and interesting planet, very different from Earth with very different phenomena. And that was sort of the Mars as we knew it as of *Pathfinder* and the early *Mars Global Surveyor*. A very desolate, but in its way, a very beautiful world that's quite interesting and quite different from the Earth. So that's my Mars. The next Mars, with more knowledge, well, that will be even harder.

Hartwell: Taking the panel topic seriously as an editor, I thought about it and said okay, well, you know if I were sitting around with a dozen science fiction writers late at night and there was a variety of them in a hotel room sitting on two double beds and that sort of thing, and some of them were hard science fiction writers and some of them were not, but they all knew one another and they were all joking and insulting each other, what kind of ideas about Martian stories would we throw around as possibilities to write now? And I came up with a short mental list which I will share with you.

[Firstly,] some kind of ancient Martian setting which is so far back that it does not relate on a one-to-one basis with what we know of Mars now, except only tangentially. Some speculation is what it might have been like in a very, very ancient time when it was cooling, but not cooled off yet, when there were those briny salt things whatever they are, and where there might have been some form of life, or maybe some form of colonization from elsewhere or, God knows [what], but it's a setting that could be used for science fiction. It's a non-human setting; nevertheless it's a setting that could be used for science fiction.

Secondly, nothing ever goes away in science fiction. You can use the traditional Burroughs-Brackett-Bradbury Mars — which is one part actual planet and one part a metaphor for the human condition — it's still a useable setting, people can still write stories there. They're far from hard science fiction stories but they can be effective science fiction literature.

[Thirdly,] I said, okay, what else? We have to go to, do varieties of politically incorrect Mars, you know: the Mars that we like blow up and destroy in order to do something else, i.e., to gather elements, to do something valuable with it, at some part of human society, at some point in the future [that] values [it]. Or it's an accident because of some kind of bizarre war, whatever.

[Fourthly,] *the* guy in Mars, I mean, C.S. Lewis, came up with this really, but there's possibilities for that; the Mars that has a kind of gestalt spirit or something.

[Fifthly,] the altered Mars. Now Geoff was talking about, you know, brown Mars, but, you know, I was thinking, well, there are lots of ways that we could alter Mars without terraforming it. Come up with purposes, okay, maybe we wanna grow mold on Mars, you know? What would we do in order to grow valuable molds on Mars, molds to which now all human beings are allergic, whatever, psychedelic molds, anything? There's lots of possibilities for stories set like that, and they would have to be grown either in a space environment or on another planet. It might be cheaper and better or in some way easier to grow them on Mars.

Moderator: Or less dangerous to the Earth?

Hartwell: Or less dangerous to the Earth, or less dangerous to human beings, whatever. And finally I think that the *Man Plus* Mars [written by Frederik Pohl, in which the protagonist is radically altered and even emasculated] has been woefully underexploited. Now I know, because I'm in science fiction, at least a dozen people who are transsexuals. I've talked to people who have gone through transsexual conversions and it's not pleasant and it's not fun, yet they do it because they feel that they must for health and other reasons. I know people who want to get rid of their bodies and get their consciousnesses downloaded into machines; I am personally acquainted with some of these people; I think they're nuts, 'cause I don't, okay, it's a different [thing for me], it's because of the way my mind works and my metabolism, versus somebody else's, you know, my sex is not somebody else's sex, okay, and everybody is entitled to volunteer, you know, and there would be volunteers, there *would* be volunteers and this would create a very, very different society, civilization, surround.

That's the farthest out I've been able to imagine as a setting, but those are possibilities for stories. Science fiction [with challenges presented to] 10 or 15 science fiction authors today, I'm sure we could come up with a few more, but that's what I could do with it, the whole range, and I welcome comments and arguments.

Moderator: I don't know if you've heard of this new, well, I don't know if it's new, mundane science fiction, I don't really understand mundane science fiction, except it doesn't sound like science fiction to me, it sounds like [the] technology [of] mundane science fiction doesn't really go along very well with spacecraft and I'm wondering [if this sort of science fiction could use Mars].

Niven: I've got a title for a Mars novel, based on this suggestion. It's titled: *Dead Mars....*

Hartwell: Takes a certain kind of writer to write that story.

Moderator: The question to be asked is, is the Mars that we know from space exploration or the Mars of past science fiction the most promising setting for science fiction in 2008 and beyond, or are we going to go elsewhere, is there another alternative?

Niven: I've been jotting notes as we go, that is for these past few days [of the Eaton Conference], and first off, I got an idea.... The idea is, it's a Draco Tavern story. Draco Tavern stories are ... constantly being visited by the Chirpsithra.... In the story, it turns out that Europeans have landed on Mars and done some work there and are claiming it. They want the Chirpsithra to decide between their claim and ours, and our claim is a bunch of little witches all over Mars, and no human being has ever been there at all.

A second idea, far future Mars — there's a whole spectrum of ideas for this — I don't think we'll ever run out. It's a far future Mars, terraformed for human use, sometimes a whole lot like Burroughs. If you like time travel you can visit Burroughs' Mars, you just have to go way the hell forward.

Third story: I'm working on this; Proxima Centauri has been building a Dyson Sphere and then they build another, and now they're in Sol system. They've taken Jupiter apart and now they're starting on Mars.

Fourth: This is, I don't know how good this notion is — it's only a few *minutes* old. Consider junk DNA; maybe 80 or 90 percent of the DNA in our genes doesn't appear to be useful for anything. What if we evolved from Martian bacteria brought by meteors? If we did that how much of that junk DNA, the inactive DNA, is geared for Mars rather than Earth? It does strike me, we could be a while finding out. The thing is wherever we go — Moon, Mars — wherever we go in the universe it's very natural for us to protect ourselves from the environment. We make a little Earth around us. But there are a few things we need not touch about Mars, one being gravity. One being the length of the day. One being the radiation level. One being Mars dust. If these things turn on our dreams, where will we go from there?

I don't have a story, it's just a notion so far. Fifth: it's been pointed out that we can contaminate Mars by a bacteria that rides on probes. We've done a whole lot to see that probes are completely sterile, and then we turn them loose in the radiation environment in the solar system, maybe it's missing the point. But, maybe meteors could have contaminated Mars billions of years ago. It's just the reverse of the situation where Mars contaminates Earth. It takes a big meteor coming in fast to raise a plume that'll put goop into orbit around the Sun, but the bacteria would survive the impact better as they impact Mars. This stuff comes in slowly. And, I've run out.

Landis: The Mars that we're learning about from the new probes is actually getting to be very different and beginning to have a more complicated history than the Mars that we knew before. The rocks that we've been finding at the *Spirit* and *Opportunity* sites..., as I'm sure you all know, is a jarosite...

Moderator: Did you all know that? How many people knew they were jarosites?

Landis: Yea, I'm sure you're all familiar with that. At least three people raised their hand and said yes. It's a ferrosulfate. What's interesting about jarosite, of course, is that it's deposited on Earth in very acid environments, very low pH environments. The kind of type of place for present-day deposition of jarosite is in the Rio Tinto formation in Spain, which Carol Stoker is principal investigator on looking for bacteria in, and there are bacteria that live in Rio Tinto, and that's about pH 1. There [are] acid-like sites in Australia that are also that acid, but pH 1 is pretty viciously sulfuric acid. Presumably this comes from Martian volcanoes emitting sulfur dioxide, converting to sulfurous and then to sulfuric acid. So the very briny shallow lakes are possibly oceans that we're talking about, [and they] are perhaps very acid lakes, very acid oceans.

And that I think has really changed my view of Mars. There is something interesting about that because present-day Mars has a lot of basalt on it, a lot of plagioclase, a lot of rocks that really are incompatible with long-term acid weathering. So this basically says that if the wet Mars was acidic it hasn't been acidic for a very long time, because there's a lot of rocks on the surface that could not be un-weathered for as long as hundreds of thousands of years if they were exposed to acidic lakes and acidic oceans. And that was a view as of *M.E.R.* [*Mars Exploration Rover*; that is, *Spirit* and *Opportunity*]. A mere three, four years ago people had this naïve view of Mars; who would think that we were so naïve about Mars?

But the latest orbiters now are looking down with global hyperspectral cameras at much higher resolution and are seeing that in the layers below these acid sulfate layers, (and below, geologically, means older), they're finding different rocks, they're finding phyllosilicates. Now phyllosilicates are the rocks of the type of which mica and clay are examples. Now phyllosilicates cannot deposit in acid conditions; they have to be neutral or perhaps basic. So the old Mars, the old old Mars now, perhaps old enough to be the Mars that had global oceans and I have to admit global oceans on Mars are pretty controversial although I like the hypothesis. The old, warm, wet Mars was not acid; it was basic or neutral.

So, *Viking* gave us a view of Mars as a planet of global climate change. *Pathfinder*, and then *Spirit* and *Opportunity*, gave us a view of Mars of having once been warmed with acidic oceans, and the new Mars observers have been giving us a view of Mars that [is] older than that and had neutral oceans. So, and it looks like Mars once was indeed a warm and wet and Earth-like planet very likely with oceans with a thick atmosphere.

And two things happened; two really bad things you don't want to happen to your planet. One of them is Mars' core cooled and it lost its magnetic field, and when the magnetic field was lost the atmosphere slowly got stripped away by primarily the water rising into the atmosphere being disassociated by ultraviolet, the hydrogen being swept away by the solar wind and the oxygen rusting the rocks to give you that rusty planet. The other very bad thing that happened was the Tharsis-forming event. It's hard to emphasize what an incredible geological feature the Tharsis bulge is, thousands of miles across with a 26-kilometer-high volcano on it; this is an immense bulge on the planet that happens to be more or less opposite to a giant impact crater, Hellas Basin, the widest and deepest feature on Mars, and I think it's up in the running for the largest impact basin in the solar system. And of course there's a lot of speculation that these two features are related. That the giant impact feature on this side of the planet created the big volcanic bulge on that side of the planet probably by shock waves.

But those two events, the loss of the magnetic field and the loss of the atmosphere, and the Tharsis impact event, transformed the planet almost to the Mars that we know today, at least to a thin-atmosphere cold planet wild with volcanoes. But the solar system wasn't done pummeling Mars with debris, and the debris that showered down onto Mars, gave it cataclysmic greenhouse effect. So when a large impactor hits Mars, it volatilizes a lot of material from the stuff that's frozen into the permafrost and this temporarily gives Mars a thick water-rich atmosphere and a greenhouse effect. And by temporarily I mean ten, a hundred thousand years — short periods on a geological scale, but still pretty long periods. But this was the thin-atmosphere volcanic Mars, so this would be the era of transient oceans that quickly acidifies. So, transient acidic oceans.

And just the view of that planet, a planet that's mostly very cold, mostly very hostile, but gets these transient impact events that make it warm and wet and acidic for periods of tens to maybe as long as a hundred thousand years; I find just a fascinating and in its way a beautiful and savage planet, and we haven't yet seen that in science fiction. And I don't know how to put it into science fiction because the history of Mars is a history of the past; it really is not a great history of necessarily the future that we're learning about, [but] just absolutely cataclysmic events that changed planets in the past and of course that makes us think of our own planet. We like to think of our planet as being ... stable and this is a nice place to live and things are pretty much the way they always have been and it really isn't true.

Our planet has also undergone radical changes and is very arguably in the middle of a very radical change right now. I hope this is not the radical change that's going to turn us into a Venus-like planet, and it probably isn't. I think we're still too close to the ice ages to yet be triggered into Venus. But it's going to happen. Mars is the story of the past, but in our solar system Venus is the story of the future; you want to look at Earth one billion years from now, look at Venus, that's our planet 'cause the Sun is getting brighter and this is something that we really don't have technology to deal with, and when it gets bright enough the oceans boil — we become Venus. But when that happens Mars is warming up; as the Sun gets brighter eventually it's going to release all of those greenhouse gases that are buried underneath Mars; eventually Mars...

Moderator: So you're saying that's where the science fiction future is going to be?

Landis: Well...

Hartwell: The secret name is Olaf Stapledon.

Landis: Yup, exactly actually!

Hartwell: We need a Stapledonian work.

Landis: Like many other things Olaf Stapledon nailed that Mars is where the Fifth Men — the Fifth Men, right — come from because the Sun got bright and it's the place [where they could evolve]. So Venus is the past, Mars is the future.

Moderator: I sense that David has some ideas to add to this. Perhaps he'd like talk about mundane science fiction and the future of Mars.

Hartwell: Well, I'll actually answer your question ... as I understand mundane science fiction, and I've talked to Geoff Ryman [who coined the term] and others about it. It's actually not dissimilar from the cyberpunk manifesto of Bruce Sterling's back in the '80s. What they're talking about [today] is a reform of science fiction so that we are restricting our speculations to hard science, but known science. They're lowering the imaginative freedom to say we do not know any way in which we can fly human beings to the planets in the solar system in any effective way to provide colonies and that sort of thing, so that's no longer an appropriate subject for science fiction. However, we do know that there's a future in artificial intelligence, there are other pieces of science that can be extrapolated on Earth or in near space.

Moderator: Can I interrupt you for a second? This is a question.

Hartwell: Yes.

Moderator: You know I read all of this, you know, no space travel.

Hartwell: It doesn't say no space travel; it says we are limited to near space, [to] the Moon, you know.

Moderator: Okay, if you look at like the Mars Society.... I mean I've taken [Robert] Zubrin as an example, because I know his work best. He lays out plans and his followers lay out plans that are perfectly logical ways that just take money, and yet mundane science fiction seems to want to keep us on Earth...

From Audience: What about cosmic radiation?

Hartwell: Well, they take more than money, they take...

Moderator: Oh, you can live underground. No, they've taken all this into consideration.

From Audience: I'm talking about the exposure back and forth, that's the problem.

Moderator: Well, read the book, read the book [*The Case for Mars* (1996) or *How to Live on Mars* (2008)].

Hartwell: I was going to say actually the mundanes could be argued into Mars, okay; at least that's my interpretation. They're trying to be as conservative as possible, oddly enough in terms of, you know, known and closely extrapolated science but they feel that that's still a very large field of possibility for writing science fiction. There is ... it's a perfectly reasonable attitude to hold except it takes away my space opera, and so, you know, I don't want my space opera taken away.

Landis: I have to say that I might agree with the mundane science fiction people except they decided to name their movement with a word that is a dictionary synonym for boring, okay? And this is part of their plan because they say, "Oh, we shouldn't be worried about being popular," so basically they're saying "We're going to write science fiction that's boring and nobody will read."

Niven: Remember what Heinlein started with, [extrapolating] about the future as if it were he [an engineer] so you could look around and pick out details. He would write about Mars, but his Mars doesn't look like that anymore, but he was writing mundane science fiction only he'd never have used the word mundane.

...

Moderator: I'm not sure David was quite finished with that response so...

Landis: It's okay, we like to interrupt David.

Hartwell: I was going to say we like to interrupt each other. This is part of the way the conversation actually goes.

Landis: No, you ... in particular...

Hartwell: Oh, all right.

Landis: It really is a conspiracy.

Hartwell: Yeah, but you know that I can hold the idea in my mind however long it takes to come back to it.

Landis: We're doing the experiment now. Mary, did you have any other questions?

Gwido Zlatkes [a conference organizer speaking from the floor]: Can we allow questions, too?

Landis: Another interruption of David! Great idea! Perfect, I'm glad you're in on it!

Moderator: Can we hold questions until maybe like 15 minutes before the end, is that all right? Because there's one thing that I really have to ask, there's a couple of things, I mean, okay, there's one exceptional question I really want to ask.

Landis: I think we succeeded.

Moderator: [inaudible]. So I will cede to David once more.

Hartwell: Next year! Next year I will answer the question.

Landis: I think you have the mic. He has the mic.

Moderator: You have the mic. We can turn the rest of them off if you like. We can do that.

Hartwell: I refer you to an article by Geoff Ryman that appeared last year in *The New York Review of Science Fiction* [June 2007] in which he clarifies and elucidates the mundane manifesto. Basically, he thinks that it is more useful in an immediate and human sense to write literature about the very large, but still limited possibilities that we face in the next hundred or so years and that if you speculate further in time than a hundred or hundred and fifty years you begin to run into magical technologies and fantasies and that's another restriction and that's basically the final restriction as I understand it.

Moderator: There will be time for questions.

Hartwell: Right.

Moderator: And this is one thing I really want to get to, and then I've got one other question after that and then we'll open the mic to questions which I'm sure you all want to ask. There may be or may not be and I'm skipping over the question whether you think it's there or not, there may be microbial life on Mars, possibly hydrothermal vent vestiges perhaps, and there are both ethical and practical questions about exploring Mars if this is true, and Stan Robinson brought these up.

If you introduce terrestrial life-forms and they compete and destroy the forms that are there, number one there's the ethical question of destroying an environment, number two there's a practical question, which is, would we never know then about these life-forms [that] might have wonderful qualities that we would like to know about but if we out-compete them, then we're not even going to have that information *and* we won't even know if they were ones that we introduced. So I guess my question is — and it's all boiled into one — if there's evidence that there's life on Mars, and I don't know if we ever proved a negative that there wasn't, should we proceed with human exploration in fact or in fiction?

Niven: First off, I think you should give up hope for the Martian bacteria; we have. Somebody is always destroying the environment. If we refrain, if somebody refrains from destroying an environment, someone else moves in. I hope we can save the things. I hope we can save a whole lot of species ... but history suggests that we're not going to do that. Secondly, this isn't easy to advertise, it's not the whales you're trying to save on Mars, it's the pond scum we should have been finding on the canals. But third, terraforming or even visiting Mars is going to be hideously expensive; on that basis maybe the Martians are safe. Point number four is I can't stop writing stories about Mars just because there are bacteria on it.

Moderator: Think of the germs.

Hartwell: Yeah. Well, anybody who knows really anything about ecology knows that what we should really worry about is bacteria and insects, and that the bulk of life on Earth [is] not big things like us and whales.

Landis: The old Case for Mars studies, again. In the 1980s, Chris McKay had led a couple of workshops on terraforming and he was actually very prescient. He said, you know, there's going to be, if we're talking about terraforming, an opposition movement — people who say we should preserve Mars as it is, and he says the fundamental question that we're going to have to deal with is, do rocks have rights? Does Mars have some kind of right to exist as it is? And of course Kim Stanley Robinson ran with that in his *Red Mars* ... where there was a faction, a political faction — that was the Reds — that wanted to keep Mars the way it is, wanted to not terraform it. And of course the related question — I think actually that's silly: rocks do not have rights; there's no reason whatsoever to keep a rocky and hostile Mars rocky and hostile. The harder question is, do bacteria have rights, does pond scum have rights? If Mars has the Martian equivalent of pond scum, does it have a right to its existence? And I think there's a lot of people there who would argue "yes," this is an alien ecology and another planet that we're not...

Moderator: Excuse me for interrupting. There's two issues, one is [inaudible] ... but I think a more substantive issue is, do we have the right to know what the other stuff was [inaudible] what was the same and what was different and do we need to know that?

Landis: Yeah, yeah.

Hartwell: The other thing is, it's not entirely [a] silly argument, there is, you can correct me if I'm wrong, but Joan Slonczewski, who's a biophysicist [and science fiction writer] told me that there's a fair amount of evidence for some rudimentary communication among bacteria. Pardon me? [inaudible comment from audience] Lots of it, yeah, lots of evidence, so we're not talking a moot point, we're talking about something that shows certain signs of sentience.

Moderator: Before we get to the audience questions, and I got plenty more questions here I could ask, but before we get to the audience questions I *did* want to ask all of you one question, which is: do you have a Martian joke?

Hartwell: I do! I've saved this joke for years. I heard this joke when I was in grade school and I saved it until my son, who is now 32, was in the fifth grade because that was the perfect time. And I have a second family and my son is in fifth grade and before he passes on this year, I will tell it to him too. The joke is: what did the spaceman say to the 50-foot-tall Martian woman? Take me to your ladder; I'll see your leader later. That's a fifth grade joke if there ever was one!

Niven: The ship lands [on Earth] and is full of Martians. The Martians rapidly learn our language and we start associating with Martians, finding out about each other. At some point they start talking about sex ... and presently the Martian is asking [for a chance to see human sex, which he is given. Is] that it? Is there more? And one of the humans says in about nine months you get a baby, and the Martian says nine months! Why were you in such a hurry to finish?

Landis: I don't think I can top that! I think I'll defer. Mary, you're the moderator.

Moderator: You don't have [one]? I warned you! I've warned you for a week and a half and you don't have a Martian joke?

Landis: You did not! You're fantasizing!

Moderator: I warned you, seriously! All right, Geoffrey does not have a Martian joke, so does anyone in the audience have a Martian joke they'd like to give to Geoffrey? Yes!

Audience member Joe Miller: I have perhaps the most horrible pun this year, based on something that Eric [Rabkin] said about character names in Bradburyan context, and the French word *merde*. I think the proper response to mundane science fiction would have to be "merde-ane" science fiction.

Moderator: Okay, I think there's time for questions. There's Jim in the front row here, wait for the microphone.

Audience member Jim Benford: The comment about mundane science fiction — it's wrong. The proposition is wrong. We already have technology solutions [such as] real ships that is, carrier-sized ships [that] carry big payloads to do a whole lot of stuff and that requires no real [imagination].

Moderator: Oh yeah! I bet you can. Go ahead.

Audience member Darryl Mallett: Moving along to another planet or another island or another whatever it is, is sort of a modern thing because how many civilizations in the past moved along without even thinking about it because they didn't know about bacteria? Even in our own culture we don't worry about killing the bacteria in our toilet or in our mouth or anything like that, we just do it because we have to or we just do it because it's what we've done all along. Now, if you're on a ship in the ocean or a planet in space and your ship is sinking, you don't really worry about whether you're going to kill off something before you get where you're going, so that's another thing like Geoff said: if Mars is where we have to go because our planet's going to be absorbed by the Sun someday, maybe we should start thinking about just getting the hell off this planet so we survive, so that's my observation and I'm open to comments.

Audience member Jake [unknown surname]: First of all my name is Jake and I worked on the Rover project in the 1960s and '70s, including [inaudible]. I listened with interest to what you were saying and I've been traveling so I couldn't come and hear the whole conference [inaudible]. But you could not remember the technology that was [inaudible] and I will tell you why [inaudible]. Let me just read something real quick here so that you understand what I mean: if an effect applies to any individual, that effect cannot describe all those affected. If an effect describes all those affected, that effect cannot apply to an individual. That is a terrible burden for a writer, for particularly a science fiction writer because you can never completely connect [inaudible] incomplete and inconnected, or unconnected, must be your description. Without going much farther into this, I have a little handout here which says that Niven's *Footfall* [inaudible]. Let's get that panel of science fiction writers and we'll get the government to figure out how to get rid of these jokers [inaudible], so here we have the short-term solution in *Footfall* and we have the long-term solution in the *Foundation Series*. And in between lies [inaudible].

Moderator: May I ask you a question? [inaudible] Would you be willing to talk about this afterwards?

[Questioner gives moderator paper from which he was reading.]

Niven: Regarding *Footfall*. The mundane's readers have no trouble at all believing in the threat team. The only people who think we're daydreaming are the science fiction fans when we put together a threat team made of science fiction writers. It's just the most obvious thing in the world; it would have gone at the time of *Footfall* through Robert Heinlein, to Jerry Pournelle, to the rest of the team; it would go through Jerry through the Air Force to the rest of the team today. It's just the most obvious thing in the world, something threatens the Earth, you assemble science fiction writers [inaudible] who else would you assemble [inaudible]?

Audience member Greg Benford: I'll make two kinds of ... [inaudible]

I want to make a point about the whole rocks and rights, and by the way about Stan Robinson's remarks yesterday. Stan and the-rocks-have-rights people are speaking out of the Western Puritan tradition; they seem largely unaware that the vast bulk of humanity does not believe this way. The Hindu, the Islamic, Buddhist civilizations don't have any track of this argument. This is a very narrow world-view and it's also historically disproved repeatedly as others have remarked. So you've got to get outside the context of Western thinking in order to actually look at this problem with any dimension and the real issue is whether you should be cautious about studying microbial life on Mars and you can't make the assumption that merely studying it would destroy it because that's not true of the microbial life here. Most of the biosphere, biomass, is microbial and is beneath the ground. It outweighs us, all we upper-dwellers. That's undoubtedly true on Mars because there's probably nothing on the surface at all. You can't conceivably attack that anaerobic life with any colonization any more than we have destroyed the microbial life of this planet. So I think it's, first it's a, even, the issue itself is particularly as Stan presented it, is a very narrow view out of one culture — the Puritan tradition — which doesn't get a lot of good press. Second, it's a non-issue [inaudible]. You can't destroy it.

Moderator: Howard.

Audience member Howard Hendrix [conference organizer]: Thank you very much. And it's interesting that you characterized that as [inaudible]. There's a Clarke story, I think it's called "No Second Eden," that keeps popping up into my head, where accidentally [...] humans wipe out Venusian life-forms, all Venusian life-forms. But, my question was really to Geoff about the issue of the impactor short time greenhouse burst as it were. Is that possible to bring about that now or is that something so completely distant in the past that if we hit Mars with an impactor now, you're not going to get the same sort of reaction?

Landis: Yeah, I assume that it is a mechanism that probably still operates, but the very large impactors were much earlier in the solar system. The solar system has basically kind of chased out all of the big stuff that intersects planets. There doesn't seem to be evidence of recent major water activity, that is, global water activity, on Mars for a very long time, we don't really know how long that time is, but it's probably multiple billions of years, so all of this happened billions of years ago. With that said, there is some good evidence for transient water on Mars. There is evidence at least of fossil hot springs, and the fossil hot springs can't be billions of years old but possibly hundreds of millions of [years]. And people are looking very hard, trying to find present-day hot springs, and there's some evidence of hydrothermal activity, hot spring activity in present-day Mars. So there may still be liquid water sort of trickling out from underneath. So the jury's still out about Mars.

Hendrix: So the idea of a Martian spring in both senses of the word, is still possible?

Landis: It's still possible, indeed, yeah, in both senses of the word.

Hendrix: Okay. Thank you very much.

Moderator: Let's restrict this to one more question and then we'll be able to do a wrap-up.

Audience member Joseph Miller: This is for Geoff Landis. I know you're a Rover guy, but it seems to me there [are] some topographical features on Mars that would be very interesting from a paleontological view, such as the size of Valles Marineris since you've probably got a pretty large geological history there. Now I know there's been talk about the insectoid robots that MIT is working on and things like that. Do you see that as a technology that's going to be applied in the near future [...] to supplement the Rovers?

Landis: Yeah, there's a lot of interest in trying to explore the layered canyons. Selecting a landing site for Mars rover is a very interesting process and a very long and involved process. The scientists want really exciting landing sites, they want mountains and cliffs and layers and everything. In fact, when the science payload for the M.E.R. rovers was put together, the number-one science target was saying we have spectacularly good infrared instruments to distinguish layers, let's go into Valles Marineris where we see lots of layers. And the safety engineers said no. NO! Among other things, the landing system has a radar that looks at how close it is to the ground, and if you're coming down over a cliff and you say, oh, we're almost at the ground, we're almost at the ground, turn off the engines, and then as you're drifting downwind, you go over the other end of the cliff, this could be bad. And it's very hard to balance these requirements for interesting science with safety considerations. The answer is probably rovers that have a great deal more traverse abilities so that you can land in the safe spot and then drive over to the less safe spot. [inaudible comment from the audience about robots with legs] I love them, I absolutely love them. What you have to realize is that humans have about 3,000 years of going down the learning curve on wheeled vehicles. We're very good on wheeled vehicles and we have maybe five years cumulative experience on walking vehicles, they are much, much harder to design. I did a design study one time, actually, for a balloon mission, to try to go up into the transient water region, and the design study there was to land at the base of one of these slopes that has a transient water feature, where the wind blows uphill, and then deploy a balloon that drags a snake behind it so that you can be blown uphill to the transient water feature. And I thought that would have been a great idea for a mission and I wanna fly that one! It didn't really go anywhere. It was a design study that did not fly. They just considered it a lot of hot air, I guess.

Moderator: We're about out of time, it seems, I've got all these questions, but we're out of time and I think we should do a wrap-up, and I'd like to start with David, if I could.

Hartwell: Yeah, no, I...

Landis: We'd only interrupt him.

Hartwell: No, all I can say is that, well, this has actually been more interesting than I even thought. You know, good stuff here.

Landis: You stayed awake through the whole thing.

Hartwell: Hey!

Niven: Okay, as I say, I've been making notes. I noticed a thing about movies, and there have been a couple movies about trips to Mars recently. Doing the same thing *2001* did. In fact, you can divide science fiction movies up into two categories: (1) Ambitious, such as *2001*. The other, not too ambitious, but successful, good fiction, such as *Rollerball* or *Westworld*. The ambitious ones always turn transcendental at the end. You must have noticed. *2001* sticks to the hardware as far as it can go and then, suddenly, you're off on a trip. Both Mars movies stick to the hardware as far as they can go, and then suddenly you're meeting Martians, you're being led into the stars. It's gonna be that way forever. And the reason is you can never see the end of a space program. Wherever you pick up man's conquest of the universe, you can always see where you ought to be and from there you can always see where you ought to be and there's no obvious end, except extinction.

Hartwell: Mary? I'm sorry, are you done, Larry?

Niven: No.

Hartwell: Go, go, sorry.

Niven: I'm holding a current *Analog* [magazine] I just wanted to point out that utopias and dystopias can always be found on Mars, and it's going on now. The story in this case is ... okay, I can't find it, wait a minute! It's called "Tenbrook of Mars." It's by Dean McLaughlin [July–August 2008 issue]. A Martian is being interviewed on how he and his remaining survivors were able to survive a disaster of the collapse of a beanstalk. It's pretty much a good polemic ... I don't say polemics are bad. This is a polemic and it's good. On self-reliance. Civilizations that depend on people helping, on seeing what needs to be done and doing it. Third: In fiction, Mars is always dying. Civilizations that have died or are dying are a very convenient thing for an author. I point out, that's the way I wrote *Ringworld* and the reason was, I couldn't think of any other way to do it. Very convenient. You're always going to find it that way, but what about the reverse? Mars of the future. You [looking at Landis] pointed out that Mars is the future. It's a good approach and it's only been done by [Kim Stanley] Robinson in *Red, Blue, Green Mars* and [Roger] Zelazny gave it a try [in "A Rose for Ecclesiastes"]. I don't remember the details and that's all I've got. Go ahead [to Hartwell].

Hartwell: Well, after the end, I wanted to say that I tried to think of what could possibly be missing from this conference and what I came up with was *Voyage to the Red Planet* by Terry Bisson, which is certainly the funniest Mars novel published in the last 40 years. Everybody should read it.

Moderator: I have a concluding remark. The observation has [been] made that women don't write about Mars, which obviously isn't true: C.L. Moore, Pamela Sargent, you could probably add some others, okay Kage Baker, [inaudible from audience] ... you don't like dead people. Sheila Finch, absolutely. Anyway, I suppose we're ending on trivia rather than on the big ideas, which these Mars minds have certainly treated us to and I think they deserve a round of applause.

The editors would like to thank Melissa Conway in her capacity as Head of Special Collections & Archives, UCR Libraries, for having provided permission to use the transcriptions. Also a special thanks goes to those who produced these transcriptions: Heidi Hutchinson, Julia D. Ree, Dr. Patricia Smith-Hunt, and Manuel Urrizola, who volunteered dozens of hours of their own time in order to produce these transcriptions by the original deadline.

ABOUT THE CONTRIBUTORS

David Clayton was born and grew up in San Diego, where he received a B.A. in English from San Diego State University in 1964, and a Ph.D. in literature from the University of California at San Diego in 1972. From 1972 to 1977, he taught American studies at the J.-W. Goethe University in Frankfurt am Main. Since then, he has worked as adjunct faculty for the University of California at the San Diego, Santa Cruz and Riverside campuses, and at Chapman University and the University of La Verne. Since 2000, he has primarily worked as a PACE (Program for Afloat Education) instructor, teaching college courses to sailors on board ships of the United States Navy.

Robert Crossley is a professor emeritus of English at the University of Massachusetts–Boston, where he taught from 1972 to 2009. A biographer, editor, critic, and literary historian, he has published widely in the fields of utopian studies, science fiction studies, and literary pedagogy. He is the author of *Olaf Stapledon: Speaking for the Future* (1994), *H.G. Wells* (1986) and *Imagining Mars: A Literary History* (2010).

Terry Harpold is an associate professor of English, film, and media studies at the University of Florida, where he teaches courses in digital culture, literary theory, and the scientific romance (primarily Jules Verne). He is the author of *Ex-foliations: Reading Machines and the Upgrade Path* (2008). His essays on Verne have appeared in *Bulletin de la Société Jules Verne*, *IRIS*, *Revue Jules Verne*, *Science Fiction Studies*, and *Verniana*. He is writing several essays related to Verne's critical reception in Britain and the United States in the early 20th century, and a book on Verne's narrative and textual methods.

Howard V. Hendrix received his B.S. in biology from Xavier University (Ohio), and his M.A. and Ph.D. from University of California–Riverside. He is the author of novels, stories, and poems translated into many languages, most prominently the science fiction novels *Lightpaths* (1997), *Standing Wave* (1998), *Better Angels* (1999) and *Empty Cities of the Full Moon* (2001), as well as *The Labyrinth Key* (2004–05) and *Spears of God* (2006–07). He is also the author of two works of nonfiction. His story collections include *Testing Testing 1,2,3* (1990), *Mobius Highway* (2001), and *Human in the Circuit* and *Depth of Perception* (both 2011) from Borgo Press. He is at work on his next novel and teaches at California State University, Fresno.

John W. Huntington is a professor of English at the University of Illinois–Chicago, and has been teaching SF since 1972. He also teaches Renaissance literature and courses on literary criticism. He is the author of *The Logic and Fantasy of H.G. Wells and Science Fiction* (1982) and *Rationalizing Genius: Ideological Strategies in the Classic American Science Fiction Short Story* (1989). He has more recently edited H.G. Wells' 1937 novel about rumors of a Martian invasion, *Star Begotten* (2006).

Sha LaBare is a recent Ph.D. in the history of consciousness at the University of California–Santa Cruz. He describes himself as a "widgeteer and nexistentialist," as well as a "mage, scholar, and cryptoethicist" who "practices the ecology of everyday life at the nexus between science fiction and spirituality."

Victoria Lamont grew up in Edmonton and completed her B.A. in English at the University of Alberta in 1988. After graduation and a brief stint as an arts administrator, she decided to pursue graduate work, receiving her Ph.D. from the University of Alberta in 1998. She teaches American literature, critical theory, and literary criticism at the University of Waterloo. Her main areas of research and publishing are 19th- and early 20th-century popular westerns, particularly by women.

Bradford Lyau received his B.A. in history from the University of California–Berkeley and his M.A. and Ph.D. from the University of Chicago, specializing in modern European intellectual history. He has taught at various colleges and universities in California and in the Balkans. He has published articles analyzing both American and European science fiction. His book *The Anticipation Science Fiction Novelists of 1950s France: Stepchildren of Voltaire* was published by McFarland in late 2010.

Joseph D. Miller received his Ph.D. in 1979 from the University of Texas and did a postdoc in dopamine electrophysiology at the University of Texas–Southwestern until 1982, followed by a second postdoc in circadian neurobiology at the University of California–Riverside and the University of California–Davis until 1987. Since 1982 he has been involved in the Eaton Conference, contributing many critical essays to the conference volumes. A former NASA space shuttle project director (1982–1987), he is an associate professor in the Department of Cell and Neurobiology at the Keck School of Medicine at the University of Southern California.

Dianne Newell is a professor of history at the University of British Columbia. The author of numerous scholarly books, articles, and chapters, her specialty areas include Canadian social and economic history; history of technology; aboriginal women in the industrial economy; Pacific fisheries; and women's involvement in postwar science fiction — with particular emphasis on the works of Judith Merril and Leigh Brackett.

Phil Nichols teaches in the School of Design at the University of Wolverhampton, where he is course leader in video and film production. He writes that he "started out as a scientist and engineer, before slipping sideways into video editing, radio drama production, and film-making." His specialty areas include short fiction, short films, and film adaptation from other media, but he is particularly interested in the works of Ray Bradbury.

Christopher Palmer was educated at Sydney University, La Trobe University and the University of Oxford. He has published articles and essays on contemporary science fiction and on Shakespeare, especially Shakespeare and film, and the book *Philip K. Dick: Exhilaration and Terror of the Postmodern* (2003). He teaches in the program in English at La Trobe University in Melbourne and is writing a book on castaway narratives from Daniel Defoe to Terry Pratchett.

Eric S. Rabkin is Arthur F. Thurnau Professor and professor of English language and literature at the University of Michigan–Ann Arbor. He received his B.A. (English, 1967) from Cornell University and his Ph.D. (English, modern letters, 1970) from the University of Iowa. He has written, co-written, edited, or co-edited more than thirty volumes, many on science fiction topics, most recently including *Mars: A Tour of the Human Imagination* (2005). He has won the Pilgrim Award of the Science Fiction Research Association for lifetime contribution to science fiction scholarship.

Kim Stanley Robinson earned a B.A. in literature from the University of California–San Diego (1974), an M.A. in English from Boston University (1975), and a Ph.D. in English from the

University of California–San Diego (1982). His doctoral thesis, *The Novels of Philip K. Dick*, was published in 1984. He is the Hugo and Nebula Award winning author, most prominently, of the Mars trilogy (*Red Mars, Green Mars, Blue Mars*), as well as the Three Californias or Orange County trilogy, the Science in the Capital series and many other novels and short stories.

Jorge Martins Rosa is an assistant professor of communication sciences at the Universidade Nova de Lisboa, Portugal. After he earned a master's degree focused on video games, his Ph.D. in communication sciences focused on Philip K. Dick and contemporary cyberculture. He is the head researcher of the funded project Fiction and the Roots of Cyberculture.

George Slusser is a professor emeritus of comparative literature at the University of California–Riverside, where he was the founder and curator of the Eaton Collection. The author or editor of 35 books and 119 articles on American and European SF and science and literature, he initiated and coordinated the Eaton Conference through 2009. He was awarded the SFRA Pilgrim Award for lifetime achievement in 1986. In collaboration with Danièle Chatelain, Slusser has produced two translations/critical editions of early French-language SF works and several articles on SF as narrative. His book on Gregory Benford is slated to appear in spring 2011.

Ekaterina Yudina describes herself as a Russian native happily settled in California. She is a lecturer in Russian in the Comparative Literature and Foreign Languages Department at University of California–Riverside. Her research covers Russian avant-garde art, St. Petersburg myths, children's memoirs, and Russian science fiction.

INDEX

Achebe, Chinua 180
Across the Zodiac (Greg) 59
Aelita (Tolstoy) 3, 51–55, 57; film (Protazanov) 54
Agamben, Giorgio 136–37, 157, 160
Aldiss, Brian 35, 131, 137
All-Story Magazine 65
Les Allemands sur Vénus (Mas) 58
Always Coming Home (Le Guin) 158, 161
"And the Moon Be Still as Bright" (Bradbury) 5, 106–116, 179
Anderson, Michael 109, 116
Anderson, Sherwood 4, 97
"The Android and the Human" (Dick) 134, 136–137
Andromeda Nebula (Yefremov) 60
Antoniadi, Eugene 17
Aquinas, Thomas 91
Argyre, Gilles d' (aka Gerard Klein) 87, 91, 94
Arnold, Matthew vi, 6, 175, 177, 181, 183
Around the Moon (*Autour de la Lune*, Verne) 29, 34
Artificial Intelligence: AI (film) 159, 161
Asimov, Isaac 1, 37–38, 48, 71, 90, 147, 187
Astounding Science Fiction (Campbell) 38, 74, 79, 127
astrobiology 160–161, 183; see also Exobiology
Les Astronautes (Rosny) 37, 45
L'Astronomie populaire (Flammarrion) 58
"At the Bottom of a Hole" (Niven) 190
Atteberry, Brian 75, 79
"Autofac" (Dick) 134

"Un Autre Monde" (Rosny) 37, 41, 47
Les Aventures extraordinaires d'un savant russe (*The Extraordinary Adventures of a Russian Scientist*, deGraffigny and Le Faure) 58

Bachelard, Gaston 183–184
Baker, Kage 1, 205
Baker, Roy Ward 33–34
bandes dessinées 37
Barad, Karen 159–160
Barnard, E.E. 17, 146
Bazin, Andre 113
Beachhead (Williamson) 18–26
Bear, Greg 11, 18, 26, 159–160, 186–188
"The Beast-Jewell of Mars" (Brackett) 77, 79
Bellona's Bridegroom: A Romance (Genone) 59
Benford, Gregory 19, 26–27, 160, 186, 188, 201, 209
Benford, Jim 201
Bergson, Henri 88
Bernal, J.D. 42, 91
Besant, Annie 121
The Best of Stanley G. Weinbaum 37, 48, 161
The Best Science Fiction Stories—1949 (Bleiler and Dickty) 96
Bester, Alfred 32, 35
Beyond the Silent Planet (Lewis) 135; see also *Out of the Silent Planet*
Bible 101
The Big Eye (Ehrlich) 96
Billion Years to the End of the World (Strugatskys) 3, 69–70
Bisson, Terry 204
Bleiler, E.F. 96
Blue Mars (Robinson) vi, 5, 18,

35, 139, 142–145, 150, 181, 209
Boëx, Joseph-Henri (aka J.-H. Rosny) 37
Bogdanov, Alexander 1, 3, 51–55, 57, 146
bolt hole 151
Boyle, Danny (director) 159–160
Brackett, Leigh v, 1, 4, 13, 73–79, 193, 208
Bradbury, Ray vi, 1–7, 9–11, 18, 26, 32, 35, 38, 60–62, 73–74, 95–118, 147–148, 160, 165, 168–169, 171–174, 179–180, 183–187, 189, 193, 201, 208
Breit, Harvey 96–97, 103
Brin, David 160, 186
Bretonne, Restif de la 58
Broderick, Damien 157, 160
Brunner, John 154, 159–160
Bruss, B.R. 87–89, 94
Buck Rogers 165
Budburg, Moura 60
Bulgakov, Mikhail 60
Burning Tower (Niven and Pournelle) 190
Burroughs, Edgar Rice v, 1, 4, 13, 18, 26–27, 32, 34–35, 37, 57–59, 73–79, 147, 153, 160, 166, 170, 172, 180, 193–194
Burroughs, William S. 180
Burton, Charles E. 17
Butor, Michel 156–157, 160
Butterfly Effect 96

Calvinism 124, 128
Campbell, John W. 42, 74, 128
A Canticle for Leibowitz (Miller, W.M.) 99, 104, 167, 174
capitalism 57, 61, 63–64, 98, 132, 136–137
Carr, Michael H. 149

211

Cartier, Rudolph 34–35
Chardin, Teilhard de 156
Child, Lydia Maria 76, 79
Chirurgiens d'une planète (d'Argyre) 91, 94
CHOSEN (Carbon, Hydrogen, Oxygen, Sulfur, Energy, Nitrogen) hypothesis 19, 20
Citizen Kane (Welles) 120
Clans of the Alphane Moon (Dick) 132
Clarke, Arthur C. 3, 11, 18, 26, 99, 104, 172–174, 202
Clayton, David vi, 5, 118–129, 207
Clement, Hal 38
colonization 4–5, 11, 13–14, 26–27, 31, 37, 52, 61, 74–78, 82, 106–107, 109, 111, 124, 130–132, 134, 157, 159, 168–169, 178–182, 184, 190, 193, 197, 202
Commissariat à l'Energie Atomique (CEA), 89
"Commonsense" (Heinlein) 128
communism 51, 63, 123, 147
Compton, D.G. 11, 148, 168–169, 174
Conan Doyle, Arthur 81
Conant, James B. 96, 98, 104
Les Conquérants de Mars (de la Hire) 37
Les conquérants de l'univers (Richard-Bessière) 87, 94
Contento, William G. 95, 104
Conway, Melissa 205
Cook, Capt. James 46
Cooper, Brenda 190
Cooper, James Fenimore 76, 79, 122
The Core (film) 27
"The Country of the Blind" (Wells) 65
Cramer, Kathryn 191
Creation Myth (Niven and Cooper) 190
Crèvecoeur 98
Crossley, Robert vi, 2, 6, 165–173, 207
"The Crystal Egg" (Wells) 58, 72
Csicsery-Ronay, Istvan, Jr. 153, 158, 160

Dandelion Wine (Bradbury) 111
Darwin, Charles 37, 41, 43, 121
Davis, Richard Harding 74–75
The Day After Tomorrow (Emmerich) 159
"A Day in the Life of an American Journalist in the Year 2890" ("La Journée d'un journaliste américain en 2890," Verne) 30

The Decline of the West (Spengler) 51
Decoding Gender in Science Fiction (Attebery) 79
Decolonising the Mind: The Politics of Language in African Literature (wa Thiong'o) 180, 183–184
Definitely Maybe (Strugatskys) 69, 72
Defontenay, C.I. 58
de la Hire, Jean 37
Delany, Samuel R. 153–154, 158, 160
del Rey, Lester 169, 174
Descartes, René 46, 121
de Waal, Frans 157, 161
Dhalgren (Delany) 154, 160
Diamond, Jared 178
Dick, Philip K. vi, 1, 5, 13, 130–138, 148, 168, 174, 178, 208–209
Dickens, Charles 131
Dickinson, Emily 178
Dikty, T.E. 96
Dimension X (radio series) 109–110, 114–116
Disch, Thomas 134
Disney, Walt 99, 101, 166, 169, 174
The Dispossessed (Le Guin) 153
Do Androids Dream of Electric Sheep? (Dick) 5, 132, 134, 135–137
Le Docteur Oméga, aventures fantastiques de trois français dans la planète Mars (Galopin) 36, 58
Double Star (Heinlein) 118–119, 128
Douglas, Ellsworth 59
Draco Tavern (Niven) 190, 194
Druillet, Philippe (artist) 37
Dune (Herbert) 158, 166, 174
Dyson, Freeman 149
Dyson sphere 194

The Earth Is Near (Pesek) 6, 170–171, 174
"The Earth Men" (Bradbury) 100, 114
"Earth's Holocaust" (Hawthorne) 101, 104
Eastwood, Clint 75
Ecclesiastes 102
"Ecopoesis" (Landis) 191–192
Edison's Conquest of Mars (Serviss) 35, 59
Ehrlich, Jack 96
Eliot, T.S. 183
Elliott, Kamilla 110, 112–113, 122
Ellison, Harlan 110, 116
Emerson, Ralph Waldo 124, 128

Emmerich, Roland (director) 159, 161
"The Enchantress of Venus" (Brackett) 75, 79
Engineer Menni (Bogdanov) 52
Envy (*Zavisti,* Olesha) 60
L'Épopée martienne (Varlet) 37
L'Ère des Biocybs (Guieu) 91, 94
L'Étonnant voyage d'Hareton Ironcastle (Rosny) 42
exobiology 19, 26; *see also* astrobiology
Extraordinary Voyages (*Voyages extraordinaires,* Verne) 29, 32
extremophile vi, 6, 19–20, 24, 26, 155–156, 175, 182

Face the Flag (*Face au drapeau,* Verne) 29
Fahrenheit 451 (Bradbury) vi, 4, 95, 102–103, 111, 116, 188; film (Truffaut) 110
"The Fall of the House of Usher" (Poe) 97, 104
"Falling onto Mars" (Landis) 191
fantasy 4, 32, 54–55, 60, 73, 74, 78–79, 84, 94, 97–99, 125, 128, 130, 137–138, 151, 153, 166, 168, 170, 173–174, 191, 207
Farewell, Earth's Bliss (Compton) 11, 148, 168–169, 174
Farewell Summer (Bradbury) 99
Farmer, Philip Jose 73
Farnham's Freehold (Heinlein) 122
Faulkner, William 125
Fifty Degrees Below (Robinson) 142–143, 145
La Fin du monde (*The End of the World,* Flammarion) 58
The Final Circle of Paradise (*Khischchnye veshchi veka,* Strugatskys) 66
Finch, Sheila 205
"First Contact" (Leinster) 60
A Fistful of Dollars (Leone) 75
Five Weeks in a Balloon (*Cinq semaines en ballon,* Verne) 29
fix-up 114
Flammarion, Camille 34, 36–37, 58–60
Fleischer, Richard (director) 159
Footfall (Niven and Pournelle) 201
For a Few Dollars More (Leone) 75
Forest, Jean-Claude (artist) 37

Fort, Charles 90
Foundation series (Asimov) 201–202
France in the Age of the Scientific State (Gipin) 94
Frankenstein (Shelley) 93, 100, 104
Franklin, H. Bruce 122–124, 126–128
Franko, Carol 142, 145
Fresnadillo, Juan Carlos (director) 159, 161
From the Dust Returned (Bradbury) 112, 116
From the Earth to the Moon (*De la Terre à la Lune*, Verne) 29, 34
Frontera (Shiner) 18, 26
The Future as Nightmare: H.G. Wells and the Anti-utopians (Hillegas) 85

Gaia theory 154–155, 159
Galilei, Galileo 6, 86, 121, 150
Gallet, G.H. 37
Galopin, Arnould 36, 58
Galvan, Jill 136–137
The Game-Players of Titan 134, 136
Genesis 100
"Genesis" (Turner) 150
Genone, Hudor 59
Gernsback, Hugo 123, 187
Gerrold, David 2
Gibson, Mel 188
Gibson, William 157
Gilpin, Robert 87, 92, 94
The Ginger Star (Brackett) 77, 79
The Gods of Mars (Burroughs) 35, 73, 75, 79
The Gold Coast (Robinson) 144–145
The Golden Man (Dick) 134
The Good, the Bad, and the Ugly (Leone) 75
Graffigny, Henry de 37
Green Mars (Robinson) 5, 18, 35, 139, 141, 145, 181, 204, 209
"The Green Morning" (Bradbury) 97, 115
Greene, Robert 5, 120, 121, 127, 128
Greg, Percy 59
Greimas, Algirdas Julien 133, 135–137
La Guerre des vampires (Le Rouge) 37
Guieu, Jimmy 87, 90–91, 94
"The Gun" (Dick) 130

Un Habitant de la planète Mars (de Parville) 36, 58
Haeckel, Ernst 154

Haldane, J.B.S. 155, 161
Haraway, Donna 153–155, 157–161
Hard to Be a God (Strugatskys) 3, 72
Harpold, Terry v, 2–3, 29–35, 152, 157, 207
Hartwell, David vi, 7, 152, 160, 190–191, 193–195, 197–201, 203–205
Hatch, John 56
Hawks, Howard 120, 128
Hawthorne, Nathaniel 101–102, 104
The Heart of a Dog (*Sobach'e serdtse*, Bulgakov) 60
Heart of the Serpent (*Serdtse zmei*, Yefremov) 60, 71
Hector Servadac (Verne) 30–31
Hegel, G.W.F. 123, 135, 157
Heidegger, Martin 158
Heinlein, Robert A. vi, 3, 5, 13, 18, 26, 36, 45–48, 60–61, 118–129, 133, 135, 137, 166–168, 191, 198, 202
Helmreich, Stefan 160–161
Hemingway, Ernest 110
Hendrix, Howard V. v–vi, 1–7, 9–14, 152, 155, 159, 175–184, 202, 207
Herbert, Frank 158, 166, 174
Hillegas, Mark 84–85
Histoire de la SF moderne: 1911–1984 (Sadoul) 48
A History of the Twentieth Century, with Illustrations (Robinson) 145
Hobomok (Child) 76, 79
The Hounds of Skaith (Brackett) 77, 79
Huckleberry Finn (Twain) 97, 104
The Humanoids (Williamson) 96
Huntington, John W. v, 4, 12, 80–85, 174, 207
Hutchinson, Heidi 205
Huxley, Aldous 60
Huygens, Christiaan 1

I, Robot (Asimov) 90
"I Sing the Body Electric" (Bradbury) 111
Icehenge (Robinson) vi, 5, 139–145
Index to Science Fiction Anthologies and Collections (Contento) 104
Institute for Blood Transfusion 3, 54
Invasion of the Body Snatchers (Siegel) 66
The Invisible Man (Wells) 60, 69–72; film (Whale) 60, 70

Ioakimidis, Demetre 92, 94
The Island of Doctor Moreau (Wells) 60

Jameson, Frederic 136–137, 153, 161
Je reviens de... (Kemmel) 89, 94
J'ecoute l'univers (Limat) 90, 94
John Carter of Mars (Disney/Pixar film) 1
Johnny Appleseed 97, 147
Jones, Gwyneth 158, 161
Journey to the Center of the Earth 27
Jünger, Ernst 135

Kagarlitski, Julius 60
Kaplan, Amy 74, 79
Keir, Andrew 33
Kemmel 87, 89
Khrushchev, Nikita 56
King, Stephen 110
Kinoy, Ernest 109
Klein Gerard 87, 91–93
Kneale, Nigel 33–35
Koyré, Alexander 121, 129
Krutch, Joseph Wood 106–107, 112, 114, 116

LaBare, Sha vi, 6, 10, 152–161, 208
Lamont, Victoria v, 4, 73–79, 208
Landis, Geoffrey vi, 2, 7, 11, 19, 190–205
Lang, Fritz 57
Lasswitz, Kurd 146
Last and First Men (Stapledon) 6, 12, 47, 82, 85, 127, 155, 161, 177, 181, 184
Last of the Mohicans (Cooper) 76, 79
Leeuwenhoek, Anton van 86
Le Faure, Georges 58
The Left Hand of Darkness (Le Guin) 153, 161
LeGuin, Ursula K. 153, 158, 161
Leinster, Murray 60
Leland, Samuel Phelps 172, 180
Lem, Stanislaw 39, 157, 161
Leone, Sergio 75
Leonidas 52
Le Rouge, Gustave 36, 58
Level 7 (Roshwald) 167, 174
"Leviathan '99" (Bradbury) 111
Lewis, C.S. 4, 12, 18, 26, 80, 83–85, 135, 137, 181, 193
Lewis, R.W.B. 124
Ley, Willy 146
The Life and Thought of H.G. Wells (Kagarlitski) 60
Life on Mars: Tales from the New Frontier (Strahan) 1
Light in August (Faulkner) 125

Limat, Maurice 87, 90, 94
Littérature française martienne de 1865 à 1958, ou le merveilleux scientifique à l'assaut de la planète rouge (online bibliography) 36
London Daily Mail 30
"The Long Years" (Bradbury) 99–100, 115
The Lost Hieroglyph: A Brackett and Burroughs Adventure (game) 1
Lovelock, James 154–155
Lowell, Percival 1, 17, 26, 34–36, 146–147, 156, 160, 165–166, 169–172, 174, 176–177, 181, 184
"The Lucky Strike" (Robinson) 142, 145
Lyau, Bradford vi, 4, 86–94, 208

Made in Mars (Stork) 47
Majors, Charles 75, 79
Mallett, Daryl 201
The Mammoth e-Book of Mindblowing Mars SF 1
"Man, Android and Machine" (Dick) 136–137
The Man in the High Castle (Dick) 136
Man Plus (Pohl) 10, 26, 148, 159, 181, 185, 194
The Man Who Fell to Earth (Tevis) 166–167. 174
The Man Who Japed (Dick) 133, 137
Mapping Mars: Science, Imagination, and the Birth of a World (Oliver) 174
Mardi (Melville) 99–100, 102, 104
Margulis, Lynn 155, 159, 161
Mariner missions 1, 2, 6, 18, 26, 36, 78, 148, 165, 168, 170–173
Marooned on Mars (Del Rey) 169, 174
"Mars and Beyond" (Disney documentary) 166, 174
Mars and the Mind of Man (conference) 6, 173
Mars Crossing (Landis) 11, 19, 190, 192, 193
"Mars Is Heaven" (Bradbury) 5, 95–96, 102–103, 108, 114
The Mars Society 182, 190, 197
Mars Trilogy (Robinson) 5, 6, 18, 32, 78, 139, 141, 143–144, 209
"The Martian" (Bradbury) 99
The Martian Child (Gerrold) 2
The Martian Chronicles (Bradbury) vi, 1–2, 4–6, 9, 11, 18, 32, 35, 60, 95–98, 100–101,
103–107, 109–111, 113–116, 147, 151, 171, 174, 178–181, 184–185, 187–188
The Martian Inca (Watson) 18, 26
A Martian Odyssey (Weinbaum) 18, 26, 38, 155
Martian Quest: The Early Brackett (Brackett; Moorcock) 73, 79
The Martian Race (Benford) 19
Martian Time-Slip (Dick) 131–132, 134–135, 137, 148, 168, 174
Mas, André 58
"The Masque of the Red Death" (Poe) 97
Matheson, Richard 109–11, 114, 116
McAuley Paul 19, 27
McDonough, Mike 114, 116
McKay, Chris 151, 200
McLaughlin, Dean 204
Melville, Herman 4, 99, 104
Memoirs of My Nervous Illness (Schreber) 125, 129
Metropolis (Lang) 57
Michelangelo 124
"Micromegas" (Voltaire) vi, 4, 86–94
Miller, Joseph D. v, 2, 10, 17–28, 201, 203, 208
Miller, Walter M. 99
"The Million-Year Picnic" (Bradbury) 98, 103, 114
Mission of Gravity (Clement) 38
Mizauld, Antoine 120
Moi, un robot (Limat) 90, 94
Moby-Dick (Melville) 99, 104
The Modern Tempter (Krutch) 106, 116
Monnet, Jean 88
Moorcock, Michael 73, 79
Moore, C.L. 205
La Mort de la Terre (Rosny) 3, 36–37, 41–43
Movement Républicain Populaire (MRP) 90
Moving Mars (Bear) 18, 26, 159–160
Mundane SF 135, 194, 197–199, 201
Murray, Bruce 172, 174

The Narrative of Arthur Gordon Pym (Poe) 93
Les Navigateurs de l'infini (Rosny) 3, 36–38, 42–45, 47, 161
Neoplatonism 121
New Martian Chronicles 51
Newell, Dianne v, 4, 73–79, 208
Newton, Isaac 86,

Nichols, Phil vi, 4–5, 106–117, 208
Nietzsche, Friedrich 124, 128, 176
"Night Meeting" (Bradbury) 99, 115, 180, 182–183
"The Nine Billion Names of God" (Clarke) 99, 104
Nineteen Eighty-Four (Orwell) 13
Niven, Larry vi, 7, 27, 190–205
Noon: 22d Century (Strugatskys) 3, 62–64, 72
Nous les Martiens (Guieu) 90, 94

Oedipus 100, 141
"The Off Season" (Bradbury) 98, 115
Olesha, Yuri 60
Oliver, Morton 166, 174
On the Beach (Shute) 167, 174
Opportunity rover 32, 191, 195–196
Orwell, George 13, 60
Out of the Silent Planet (Lewis) 12, 18, 80, 83, 84, 85, 137

Pagetti, Carlo 130–132, 137
Pal, George 47, 60, 69
Palmer, Christopher vi, 5, 10, 132, 137, 139–145, 208
Parville, François Peudefer de 36
Parville, Henri de 36, 58
Pascal, Blaise 3, 46
The Past Through Tomorrow (Heinlein) 128
"The Pedestrian" (Bradbury) 97–98, 101, 103
People of the Talisman (Brackett) 77, 79
Pesek, Ludek 6, 170–171, 174
Pharaoh's Broker (Douglas) 59
Phoenix mission 22–23, 26, 183
Pickering, W.H. 17
Pierce, Hazel 135–136, 138
Pinter, Harold 110
The Pioneers (Cooper) 122
"Piper in the Woods" (Dick) 130
Planet Stories 73, 75, 79, 95
Planetary Society 187
La Planète Mars (*The Planet Mars*, Flammarion) 36, 58
Planetomachia (Greene) 5, 120, 127, 128
Plato 100, 104
Podkayne of Mars (Heinlein) 118–119, 128
Poe, Edgar Allan 4, 93, 97, 104
Pohl, Frederik vi, 7, 10, 19, 26, 148, 159, 181, 185–189, 194
Pontano, Giovanni 120
Les Posthumes (de la Bretonne) 58

Index

Pournelle, Jerry 47, 190, 202
Princess of Mars (Burroughs) 18, 45, 47, 57, 73–74, 79
Prisoners of Power (Strugatskys) 67
Le Prisonnier de la planète Mars (Le Rouge) 37, 58
Propp, Vladimir 133
Protazanov, Yakov 57
"providential grace" 31, 35

Quatermass and the Pit (Kneale) 33–35
"Queen of the Martian Catacombs" (Brackett) 75, 79

Rabkin, Eric S. v–vi, 1–7, 10, 95–104, 116, 138, 201, 208
Rainbow Mars (Niven) 190–191
Rains, Claude 70
Rayon Fantastique 37, 45
Reagan, Ronald 190
The Reavers of Skaith (Brackett) 77, 79
Red Mars (Robinson) 5, 18, 34–35, 141–142, 144–145, 160–161, 171, 174, 181, 191, 193, 200, 209
Red Planet (Heinlein) 18, 60, 118–119, 123, 126, 128
Red Star (Bogdanov) 3, 51–55, 57, 146
Ree, Julia D. 205
Remaking History (Robinson) 142, 145
"Remarks Occasioned by Dr. Plank's Essay 'Quixote's Mills'" (Lem) 157, 161
Revelation 102
Rêves étoiles (Flammarion) 36
Richard-Bessiere, F. 87, 94
"Ridge Running" (Robinson) 142, 145
Ringworld (Niven) 204
Ringworld's Children (Niven) 190
Roadside Picnic (Strugatskys) 3, 68, 72
Robida, Albert (artist) 58
Robinson, Kim Stanley vi, 2, 5–6, 10, 12–13, 18, 26, 32, 35, 78, 130–152, 156, 160–161, 171, 174, 181, 186, 199, 200, 202, 204, 208
Robur the Conqueror (*Robur-le-Conquérant,* Verne) 29
Rocket Ship Galileo (Heinlein) 58
"Rocket Summer" (Bradbury) 98, 100, 114, 185
Rollerball 204
Rosa, Jorge Martins vi, 5, 130–138, 209

"A Rose for Ecclesiastes" (Zelazny) 18, 204
Roshwald, Mordecai 167, 174
Rosny, J.H. v, 1, 3, 10, 36–48, 159, 161
Ryman, Geoff 197, 199

Sadoul, Jacques 45, 48
Sagan, Carl 20, 149, 172–174, 192
Saint Paul 100
Saint-Simon, Comte de 87–89
The Sands of Mars (Clarke) 11, 18
Sargent, Pamela 205
Sarris, Andrew 120, 129
Scatterbrain (Niven) 190
Schiaparelli, Giovanni 2, 17, 34, 59, 65, 165–166, 170, 176
Scholes, Robert 137–138
Schreber, Daniel Paul 125, 129
SDI (Strategic Defense Initiative) 190
Sebeok, Thomas A. 158
The Second Invasion from Mars (Strugatskys) 65–67, 72
The Secret of Life (McAuley) 19, 27
"A Sensitive Dependence on Initial Conditions" (Robinson) 142, 145
Serling, Rod 110
Serres, Michel 31, 35
Serviss, Garrett P. 34–35, 59
"The Settlers" (Bradbury) 99, 115
Seznec, Jean 120–121, 129
Shelley, Barbara 33
Shelley, Mary 93, 104
Shiner, Lewis 18, 26
"A Short Sharp Shock" (Robinson) 141, 143, 145
Shute, Nevil 167, 174
Siegel, Don 66
"The Silent Towns" (Bradbury) 99, 115
The Simulacra (Dick) 5, 132–137
Sixty Days and Counting (Robinson) 143, 145
Slonczewski, Joan 200
Slusser, George v, vi, 1–7, 10, 36–48, 56–72, 78, 116, 138, 152, 159, 185, 209
Smith-Hunt, Patricia 205
Snyder, Gary 150
Solar Lottery (Dick) 130, 133, 136
Solaris (Lem) 39
Soldiers of Fortune (Davis) 75, 76
Solzhenitsyn, Alexander 71
Something Wicked This Way Comes (Bradbury) 111

"Somewhere a Band Is Playing" (Bradbury) 111
S.O.S Soucopes (Bruss) 94
"A Sound of Thunder" (Bradbury) 96, 103, 108
Soylent Green (Fleischer film) 159, 161
Space Apprentice (Strugatskys) 66. 72
Spengler, Oswald 51
Spenser, Edmund 13, 185
Spielberg, Steven 159, 161
Spirit rover 32, 191, 195–196
Stableford, Brian 73, 74, 78, 79, 153
"Stanzas on the Grand Chartreuse" (Arnold) 175, 184
Stapledon, Olaf 6, 12, 42, 47, 80, 82–85, 127, 155, 159, 161, 177–178, 181, 183, 197, 207
The Star Begotten (Wells) 4, 61, 8–85, 167, 174, 207
Star, ou psi de Cassiopée (Defontenay) 58
Star Trek 60, 158
Star Wars (films) 154
"Star Wars" 190; *see also* SDI
Starman Jones (Heinlein) 61
The Stars My Destination (Bester) 32, 35
Starship Troopers (Heinlein) 122
Sterling, Bruce 197
Stewart, Patricia 1
Stites, Richard 54–55
Stoker, Carol 195
Stork, Christopher (aka Stéphane Jourat) 47
Strahan, Jonathan 1
Stranger in a Strange Land (Heinlein) 5, 18, 45, 118–119, 121–125, 127–128, 133, 135, 137, 166–167, 174
Strugatsky, Arkady v, 1, 3, 56–57, 59, 61–69, 71–72
Strugatsky, Boris v, 1, 3, 56–57, 59, 61–69, 71–72
Sturgeon, Theodore 186
Sullivan, Walter 172–174
"The Summer Night" (Bradbury) 100, 114
Sun Ra 159
Suvin, Darko 137, 147
Swift, Jonathan 133

"Talk with Mr. Bradbury" (Breit) 96–97, 103
Taylorism 54, 59
TEGA (Thermal and Evolved Gas Analyzer) 23–24
"Tenbrook of Mars" (McLaughlin) 204
Les Terres du ciel (*The Earths of the Sky,* Flammarion) 58
Tevis, Walter 166–168, 174

"There Will Come Soft Rains" (Bradbury) 96, 98, 108, 115
The Thing from Another World (Hawks) 120, 128
Things Fall Apart (Achebe) 180
Things to Come 188–189
"The Third Expedition" (Bradbury) 5, 98–99, 1-2-103, 107, 112, 114
The Three Stigmata of Palmer Eldritch (Dick) 131, 136–137, 148, 168, 174
Time Enough for Love (Heinlein) 127
The Time Machine (Wells) 47, 60
Time Out of Joint (Dick) 136, 137
Les Titans du ciel (*The Titans of the Sky*, Varlet) 59
Tolstoy, Alexei 3, 51–55, 57, 67
"Tomorrow's World" (Del Rey) 169
Topsy-Turvy (*Sans dessus dessous*, Verne) 30
Total Recall (film) 147
"The Translator" (Robinson) 142, 145
Truffaut, François 110
Turner, Frederick 150
Turzillo, Mary vi, 1, 7, 190
Twain, Mark 4, 97, 104
28 Days Later (Boyle film) 159–160
28 Weeks Later (Fresnadillo film) 159, 161
"Twilight" (Campbell) 42
Two Planets (*Zwei Planeten*, Lasswitz) 146
2001: A Space Odyssey (Clarke and Kubrick) 204

Ugly Swans (Strugatskys) 3, 67–68, 72
"Universe" (Heinlein) 128

The Unteleported Man (Dick) 130
Uranie (Flammarion) 58
Urrizola, Manuel 205
"Usher II" (Bradbury) 97, 115

VALIS (Dick) 137
Valles Marineris 25, 27, 203
Valley of Dreams (Weinbaum) 155
Van Vogt, A.E. 114
Varlet, Théo 37, 59
"The Veldt" (Bradbury) 108
Verne, Jules v, 1–3, 27, 29–38, 58, 81, 87, 157, 160, 207
Viking missions 18, 21–23, 26–28, 78, 142, 148–149, 173, 181, 196
Vint, Sherryl 158
Voltaire 1, 4, 45, 86–94, 208
Voltaire's Micromégas: A Study in the Science, Myth, Art (Wade) 94
von Braun, Wernher 146, 166
von Uexkull, Jakob 158
"Le Voyage" (Rosny) 42
Voyage de cinq américains dans les planètes (de Graffigny) 37
Voyage to the Red Planet (Bisson) 204

Wade, Ira O. 86–87, 94
Wallace, Alfred Russell 17, 160
The War of the Worlds (Wells) 1, 4, 6, 18, 26, 32, 35, 56, 58–61, 63–64, 66, 69–71, 81, 118–119, 121, 129, 147, 154–155, 159, 161, 167–168, 174, 176–178, 181, 183–184
"War Veteran" (Dick) 137
The Warlord of Mars (Burroughs) 35, 75, 79
wa Thiong'o, Ngugi 180, 183–184
Watson, Ian 18, 26

"Way in the Middle of the Air" (Bradbury) 97, 115
We (Zamyatin) 13, 59, 72
"We Can Build You" (Dick) 130, 137
"We Can Remember It for You Wholesale" (Dick) 168, 174
Weiler, A.H. 96, 104
Weinbaum, Stanley G. 1, 18, 26, 37–38, 45, 48, 155, 159, 161
Welles, Orson 56, 120, 147
Wells, H.G. v, 1–4, 6, 11, 13, 17, 26, 30, 32–33, 35–37, 47–48, 56–73, 80–85, 89, 118–121, 129, 143, 147, 149, 154–155, 158–159, 161, 167–170, 174, 176–181, 183–184, 207
Westworld 204
Whale, James 60, 69
When Knighthood Was in Flower (Majors) 75, 79
Whitman, Walt 128
"Who Goes There?" (Campbell) 128
Williamson, Jack 18, 26, 96
Winesburg, Ohio (Anderson) 97, 99, 101, 103
Wollheim, Donald 96, 104
The World Jones Made (Dick) 133, 135–137

Les Xipéhuz (Rosny) 37, 41

Yefremov, Ivan 60–62, 64, 71–72
Young, Charles A. 17
Yudina, Ekaterina v, 3, 51–55, 208

Zamyatin, Evgeny 13, 59–60, 72
Zelazny, Roger 18, 204
Zlatkes, Gwido 198
Zubrin, Robert 190, 197

www.ingramcontent.com/pod-product-compliance
Ingram Content Group UK Ltd.
Pitfield, Milton Keynes, MK11 3LW, UK
UKHW050528150426
5217IPUK00026B/1851